D0849975

PRAISE FOR

HOW INNOVATION REALLY WORKS

"Every corporate leader gives lip service to the need for innovation as the engine of organic growth, but how many actually measure the productivity of their innovation strategies and expenditures? In *How Innovation Really Works*, Anne Marie Knott provides a truly new and eminently practicable way of measuring and thinking about the productivity of R&D. For everyone who manages or thinks about management, Anne Marie Knott's work is a living illustration of the clarity that flows from having good measures and benchmarks. If you want to see how your company measures up or learn how to improve your innovation batting average, get this brilliant book now."

—RICHARD RUMELT

the Harry and Elsa Kunin Chair at the UCLA Anderson School of Management and author of *Good Strategy/Bad Strategy*

"*How Innovation Really Works* will rock your world. Knott goes right to the core R&D dilemma: declining innovation in the face of soaring investment. By deeply probing inconsistencies between theory and reality, she shines a bright light on common prescriptions for superior R&D that simply don't work—and provides solutions that do. A superb read for those who care passionately about the next innovations."

—KATHLEEN M. EISENHARDT

S.W. Ascherman Professor and codirector of the Stanford Technology Ventures Program at Stanford University, and coauthor of *Simple Rules*

"This treasure chest of insight is a must-read for anyone interested in innovation—managers, investors, and researchers alike. Anne Marie Knott's RQ measure of innovation effectiveness is an invaluable compass for navigating through the fog of R&D investment, returns, and organization."

—RON ADNER

the David T. McLaughlin Chaired Professor at the Tuck School of Business of Dartmouth College and author of *The Wide Lens*

"RQ is an invaluable tool for CTOs and those involved with leading and funding technology development and innovation. Knott provides astute insights and a quantitative tool to define and improve the productivity of innovation spending, inform strategic discussions on optimizing investment at a company level, as well as how to best optimize among divisions."

—BRUCE BROWN
former CTO of Procter & Gamble

"When it comes to innovation, many companies are like Alice in Wonderland—they don't know where they want to go and so they can't choose which road to follow. Knott provides a simple but powerful signpost that leaders can use both to evaluate their prior innovation decisions and to help make future innovation decisions more effectively."

—JAY BARNEY
Lassonde Chair of Social Entrepreneurship at the David Eccles School of Business at the University of Utah and author of *What I Didn't Learn in Business School*

"Anne Marie's analysis challenges several myths and misconceptions on how innovation, growth, and profitability are mutually related. Her original and methodical approach, based on the introduction of an R&D investment variable in the classic production function, provides a very useful and practical tool for executive managers to set optimal R&D investment targets focusing on growth and ROI."

—ALESSANDRO PIOVACCARI
CTO of Silicon Labs

"Galileo Galilei, in describing the scientific method, said: 'measure what is measurable, and make measurable what is not so.' Professor Knott's research has made measurable and precise how companies can become more effective at research and development. This book provides a modern and comprehensive guide to innovation management and is a must-read for executives who care about improving their company's performance through breakthrough products and services."

—KARIM LAKHANI
Professor of Business Administration at Harvard Business School and author of *Revolutionizing Innovation*

"Anne Marie's RQ work is generating a great deal of buzz in the field. She has created an original, convincing, and empirically validated measure that challenges many of the assumptions and conclusions of famous papers in management and economics. Armed with this measure, this book provides corporations with powerful advice about how to elevate R&D as the engine of value creation."

—TODD ZENGER
the N. Eldon Tanner Chair in Strategy at the David Eccles School of Business at the University of Utah and author of *Beyond Competitive Advantage*

"Just like people have an IQ, companies have an RQ: a Research Quotient. With her painstaking research and data, Anne Marie Knott shows how RQ powers innovation for companies and for regions and nations as a whole. If you want to understand how innovation really works, read this book."

—RICHARD FLORIDA
Global Research Professor at New York University, University Professor at the University of Toronto, and author of *The Rise of the Creative Class*

"Getting the innovation resource allocation decision right is critical to shareholder value creation, demanding metrics to measure productivity and performance. *How Innovation Really Works* is both a thorough review of existing R&D metrics and a description of a new approach that seeks to better show the value of R&D. This thought-provoking book is a valuable contribution to the field. I very much enjoyed reading it."

—A. N. SREERAM
CTO and SVP of The Dow Chemical Company

"Billions of dollars are spent each year on R&D, with limited ability to measure the impact on the value it provides to help in proper allocation of limited resources and to optimize shareholder value. Drawing on past research and firsthand experience, as well as interviews with current practitioners in the area, *How Innovation Really Works* is a book that needed to be written."

—ROBERT W. FRICK
former Vice Chairman and CFO of Bank of America

"*How Innovation Really Works* is a refreshing and at times even controversial new look at innovation and return on investment (ROI) from the R&D activities of every major industrial enterprise. Knott uses solid scientific data to support her premise that ROI from innovation is not necessarily proportional to money spent on R&D. She clearly describes the complexity of the subject: how R&D is accounted for, how innovation impacts the outcome, and how that outcome is tied to revenues and timing. She even considers such factors as behavior and human talent. She proves that the most popular recommendations for ROI from innovation do not work and proposes a new and relatively simple measure: RQ (a sort of corporate R&D IQ), somewhat of an equivalent to human IQ."

—MIETEK GLINKOWSKI
VP of Global Engineering NA at Schneider Electric

HOW INNOVATION REALLY WORKS

ANNE MARIE KNOTT

New York Chicago San Francisco Athens London Madrid
Mexico City Milan New Delhi Singapore Sydney Toronto

1 2 3 4 5 6 7 8 9 QFR 22 21 20 19 18 17

ISBN 978-1-259-86093-5
MHID 1-259-86093-0

ISBN 978-1-259-86094-2
MHID 1-259-86094-9

DISCLAIMER: Any opinions and conclusions expressed herein are those of the author and do not necessarily represent the views of the U.S. Census Bureau. All results have been reviewed to ensure that no confidential information is disclosed.

To Nick,
may you also and always be well pleased

CONTENTS

PREFACE

My initial insight for what would become RQ™ (short for research quotient) came from an earlier career in defense electronics at Hughes Aircraft Company. Until it was acquired by General Motors (GM) in 1985, Hughes had been a marvelous company of talented scientists and engineers working on exciting projects. To give you a sense of how special the company was, I once asked my customer how his agency chose when to award contracts to Hughes versus one of our competitors. He said, "I go to Hughes when I don't know what I want. Once you help me discover that, I go to the other guys because they can build it more cheaply."

At Hughes, we didn't view our job as designing missiles; we viewed our job as pushing the knowledge frontier. So I became alarmed when the government changed its acquisition policies in ways that reduced companies' incentives to conduct R&D. I became more alarmed after the GM acquisition, when management began asking us to include return on investment (ROI) estimates on proposals for R&D projects. Prior to that time, proposals merely identified the challenges prompting the need for the project, the doors the project would open, and the resources the project would require, and when it would require them. These proposals were all submitted to senior management (most of whom were themselves scientists and engineers), as the rest of us waited for the proverbial white smoke to appear. There was no need to quantify the returns because the company ethos was "do the right thing and profits will follow." I felt moving toward ROI-based project selection was a death knell for the types of projects that had made Hughes so exciting—and successful. There was no market for communications satellites when Harold Rosen and Donald Williams originally proposed the project to Pat Hyland in 1959.[1] Hughes created that market.

I was concerned that the combination of government policy changes and company strategy changes were going to permanently degrade Hughes's ability to push the knowledge frontier, but I had no quantifiable means to demonstrate that. Because there was no good measure of R&D capability, I couldn't show that it was deteriorating.

I became an academic in part to solve that problem. Over the course of a 20-year career doing research and consulting on innovation, I devised RQ as a measure of a company's R&D capability—its ability to convert investment in innovation and R&D into products and services people want to buy or into lower cost for producing those products and services.

Once I had the RQ measure, I knew it was the holy grail I'd been seeking while at Hughes. My first step was to characterize the R&D productivity of all publicly traded U.S. companies going back as far as data would allow (1972). I learned that my concerns while at Hughes were valid—not only for Hughes but for the entire U.S. economy. The next step was to identify what caused the decline and, accordingly, how to reverse it. There was no easy way to do that because companies keep their R&D practices pretty close to their vests.

Fortunately, I was awarded two National Science Foundation (NSF) grants that allowed me to conduct ethnographic interviews with companies to understand their R&D programs and practices and to quantify the impact of these practices across the full spectrum of U.S. companies engaged in R&D. Through these two studies, I have been able to identify which popular innovation recommendations actually *do* work. The results are startling. Many of them don't work. Not only do they fail to improve innovation, in many cases they actually make companies worse at innovation!

In this book, I share these results. My goal is to diffuse the RQ measure so that it can do for R&D what Total Quality Management (TQM) did for manufacturing, what hospital report cards are doing for morbidity, and what sabermetrics is doing for baseball. RQ not only tells companies how "smart" they are, it provides a guide for how much they should invest in R&D and innovation and how much that investment will increase revenues, profits, and market value. If companies use these findings to increase their RQs, it will not only result in their own desired growth, it should also restore economic growth to the rates we enjoyed in the mid-twentieth century.

ACKNOWLEDGMENTS

The most obvious person to acknowledge in these pages is Jay Barney. About six years ago, he saw me present my work on RQ and said, "You need to write a book!" He put me in touch with Melinda Merino at Harvard Business School Publishing, who said, "This is an editor's dream!" Her enthusiasm and a tremendous amount of direction and rewriting by David Champion led to the *Harvard Business Review* article "The Trillion-Dollar R&D Fix," which forms the basis of Chapter 4.

The *HBR* article, in turn, led Ilana Golant at NBCUniversal to partner with me in creating the CNBC RQ 50 innovation ranking. I'm extremely grateful to Ilana for bringing the partnership to fruition, to David Spiegel and his team of writers, Susan Caminiti, Bob Diddlebock, Lori Ioannou, Ari Levy, and John Schoen, for generating compelling content, and to Dominic Chu for bringing it to life on air.

I'm especially grateful to Elizabeth Little, a former literary agent and now successful author of *Dear Daughter*, for spending many hours explaining the mysterious world of publishing and revising drafts of the proposal. Ultimately, the colleague who opened the door was Ron Adner, author of *The Wide Lens*, who introduced me to Esmond Harmsworth at Zachary Shuster Harmsworth Literary. Esmond was a joy to work with. He achieved a great balance of enthusiasm and penetrating insight with tough critique in forcing me to develop a more compelling book. Casey Ebro, my editor at McGraw-Hill, took over where

Esmond left off. Casey was creative, responsive, and tremendously capable in designing, editing, and managing the book. I'm grateful for all her help in bringing *How Innovation Really Works* to the world.

That's the story of the book itself, but of course there would be no book without the scholars who have taught me and worked with me over the years. They are too numerous to mention, so I focus on those who contributed directly to the book. I'm grateful to my colleagues at Olin: Nick Argyres, Lyda Bigelow, Dan Elfenbein, Bart Hamilton, Jackson Nickerson, Lamar Pierce, and Todd Zenger, for numerous brown bags where I sketched out the ideas comprising the book. I'm thankful to colleagues at academic seminars and conferences in which I've been fortunate to present portions of this work: at Columbia University, Georgia Institute of Technology, London Business School (LBS), Stanford Institute for Economic Policy Research, University of Kansas, University of South Carolina, Western Ontario University, the Sumantra Ghoshal Conference at LBS, the Darden and Cambridge Judge Entrepreneurship and Innovation Research Conference, the Organization Science Winter Conference, the Census Data Research Conference, DRUID15, International Society for New Institutional Economics (ISNIE) Meetings, the Atlanta Competitive Advantage Conference, The Strategic Research Initiative, the BYU/University of Utah Winter Strategy Conference, and the Wharton Technology Conference. I'm also grateful for comments from practicing R&D managers at the Industrial Research Institute (IRI) annual conferences and practicing investment professionals at the CFA Institute annual conference.

More important, I'm thankful for the talented colleagues who coauthored with me on papers utilizing RQ, and who therefore shaped many of the ideas in this book: Michael Cooper, Trey Cummings, Romel Mostafa, Carl Vieregger, and Wenhao Yang. Relatedly, I'm grateful to the editors—including Pankaj Ghemawat and Bruno Cassiman at *Management Science* and Connie Helfat at the *Strategic Management Journal*—and anonymous reviewers at the journals in which the work has been reviewed and/or published.

This book would not have been possible without the generous support of the National Science Foundation (NSF) studies for which *How Innovation Really Works* serves as an informal final report. Accordingly, I am grateful to Julia Lane and Nimmi Kannankutty, the program managers at NSF, for recognizing the promise of RQ and funding the two studies. I am thankful to Bruce Brown, then CTO of Procter & Gamble, and Frank Doerner, VP Boeing Research and Technology, for sponsoring in-depth studies in their respective firms, as well as Abigail Cooke and John Sullivan at the University of California Los Angeles Census Research Data Center, for their help accessing data and obtaining approvals to release results. I am also grateful to Richard Mahoney for the generous financial support of the *Olin Award: Research That Transforms Business.*

I want to thank the numerous R&D managers who have participated in interviews to help me better understand the real-world benefits of RQ, as well as the practical problems implementing it: John Reid, Marcus Parsons, Kathy Kuberka, Daniel Abramowicz, Martha Collins, Erik Antonsson, David L. Morse, Stephen Johnson, Richard Michelman, J. Stewart Witzeman, Stephen Toton, Johanna Dwyer, James Euchner, and Mark Matlock. You are the ones with the power to realize the potential of R&D to restore economic growth.

Finally, I want to thank Jeff Hirsch and Marilyn Gannon for their help over the past five years translating academic thinking on RQ into things that managers cared about, and CeCe Myers for keeping the rest of my work life in order as I pursued the book.

THE PROBLEM: Flying Blind

There is no question that innovation is important. Everywhere you turn, people are lauding its benefits. It's hard to open any popular business magazine and not find an article on innovation. This preoccupation with innovation comes from the belief that it is the key to growth. As Strategy& reported in its 2015 Global Innovation Study, "The results of our survey of 1,757 executives couldn't be clearer: innovation today is a key driver of organic growth for all companies—regardless of sector or geography."[1] Indeed, the company reports its Innovation 1000 companies (top 1000 R&D spenders) invested $680 billion in R&D last year—up 5 percent from the prior year.

The benefits of innovation don't stop with companies. Historically R&D has been viewed as the engine of economic growth as well. This assumption was the foundation for President George W. Bush's America COMPETES (Creating Opportunities to Meaningfully Promote Excellence in Technology, Education, and Science) Act of 2007, whose goal was to invest in research and development to improve the competitiveness of the United States. Demonstrating that support for innovation is nonpartisan, President Barack Obama signed into law the America COMPETES Reauthorization Act of 2010 three years later.

Yet despite the importance of innovation to companies as well as to the broader economy, despite the growth in real R&D by both the government and companies, and despite all the experts dedicated to helping companies innovate, companies have become worse at it! The money companies spend on research and development is producing fewer and fewer results. In fact, the returns to companies' R&D spending have declined 65 percent over the past three decades. Not coincidentally, this decline coincides closely with the decline in U.S. GDP growth over the past 30 years (see Figure 1-1).

Given the tremendous importance of innovation, all the attention paid to it, and all the experts dedicated to advising

FIGURE 1-1. Decline in R&D productivity coincides with decline in GDP growth

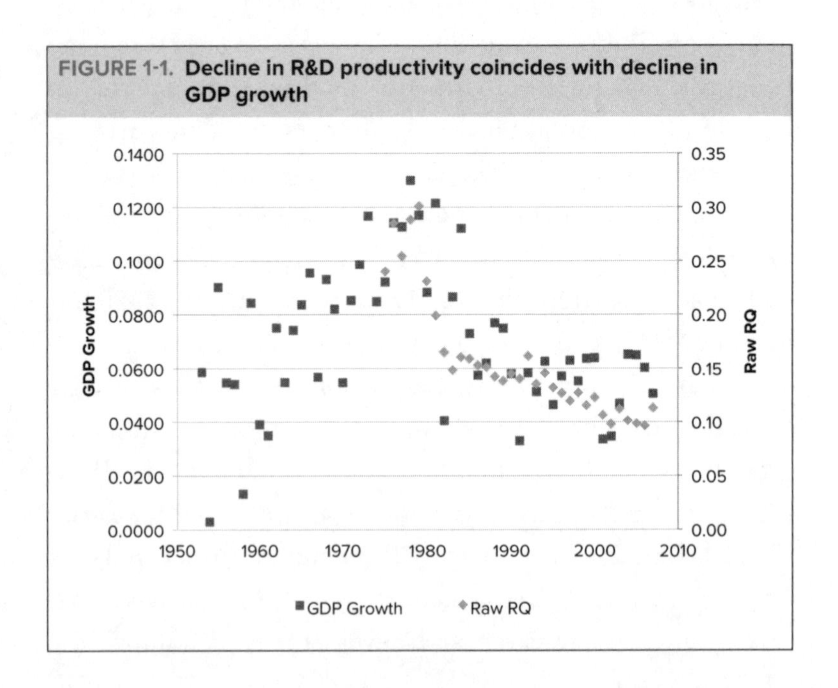

companies on it, how is it possible that R&D has suffered such a severe decline in productivity?

I believe it is because everyone is flying blind with respect to innovation, because there has been no good way to measure its quality or productivity. Indeed, Industrial Research Institute (IRI) members report that the lack of good R&D measures is one of the top problems they face.[2] They view measurement as important (a) to justify R&D investment to CEOs, the board, and investors, (b) to improve the efficiency of R&D, and (c) to estimate the value of R&D investment for future growth.

EXISTING MEASURES

This measurement problem isn't about a dearth of measures. In fact, it's the opposite problem—overabundance of measures. According to one study by the European Industrial Research Management Association, there are over 250 R&D metrics! Certainly one reason for the plethora of measures is that capturing project-level performance is quite different from capturing company-level performance. However, the more compelling reason appears to be that the measures are unsatisfactory: the data is difficult to collect, there aren't uniform standards across business units, the measures can be gamed to make sure a given group looks good, the measures focus on inputs and outputs (rather than the conversion of inputs to outputs), and perhaps most important, the measures aren't meaningful to shareholders.[3]

Until now, academics have been unable to help companies solve the measurement problem. That's because their primary measure (patent counts) wasn't much better. Patents have a number of shortcomings that are acknowledged by academics and practitioners alike. In particular, patents are not *universal*, *uniform*, or *reliable*.

The first problem, *universality*, is that not all companies patent their inventions. In fact, less than 40 percent of companies who conduct R&D have any patents. Relatedly, even among the companies that do patent, few of them patent all their innovations. This is because patents are costly both financially (the cost to file and defend) as well as competitively (they require disclosure of the fundamental knowledge underpinning the innovation). Accordingly, companies file patents only under certain circumstances, such as to prevent copying when their innovations are easy to invent around. However, they also patent for strategic reasons, such as to block other companies' patents, to prevent lawsuits, to use in negotiations with companies who hold patents to necessary technology, or to enhance their reputation.[4] Without universality (all companies patenting all innovations), it is difficult to use patents to compare companies on their innovativeness, or even to track any given company's innovativeness over time.

Patents also suffer the *uniformity* problem that they aren't all created equal. Compare for example the $2 billion in royalties for Kary Mullis's patent for the process to clone DNA[5] to the value of the 97 percent of patents that are never commercialized. On average, less than 10 percent of patents account for 80 to 85 percent of the economic value of all patents.[6] Without uniformity, the number of patents is not a meaningful measure of the value of innovations. While there are efforts to control for the uniformity problem through counting patent citations, these efforts are only partially effective, and they are only meaningful after adequate time has elapsed since the patent has been granted.

The final problem, *reliability*, is that patents don't predict the big outcomes that companies care about, such as revenues, profits, and market value. This is not surprising given the prior two problems. However, the most insightful answer to the question of why patents aren't reliable came from Dan Stern, a former vice president and chief scientist at Olin Corporation, who I run into periodically at the local happy hour. When I mentioned I was examining patents as a measure of innovativeness, Dan became very animated and said, "Patents don't measure anything! I know exactly how to increase patents—I merely tell my engineers they're going to get paid per patent."

Without *universal*, *uniform*, and *reliable* measures of innovation, it is almost impossible to identify best practices or establish top-level R&D strategy. In short, companies confront the classic problem "you can't manage what you can't measure," a quote that has been attributed to a number of people, dating back as far as 1883 to Lord Kelvin, and more recently to Peter Drucker, Fred Smith (FedEx founder), and Andy Grove (former Intel CEO). To make the "you can't manage what you can't measure" problem more concrete, let's review one example of how the problem manifests itself in the context of R&D.

In the summer of 2010, Mark Hurd was ousted as the CEO of Hewlett-Packard because there had been "violations of H.P.'s standards of business conduct." Joe Nocera, in his August 13 column in the *New York Times* that year, suggested the business conduct rationale was a ruse. Nocera cited Charles House, a former H.P. engineering manager, as saying, "the sexual harassment charge (against Mark Hurd) was a total red herring." Nocera goes on to report, "as many H.P. old-timers saw it . . . Mr. Hurd was systematically destroying what had always made H.P. great. The way H.P. made its numbers, Mr. House said, was not just by cutting any old costs, but by 'chopping R&D,' which had always been sacred at H.P. The research and development budget used to be 9 percent of revenue, Mr. House told me; now it was closer to 2 percent."

Is Mr. House right? Did Mark Hurd destroy what made HP great? The problem with existing measures is that we have no way of knowing. We don't know (1) whether R&D capability has deteriorated at all, or if so by how much, and (2) whether the correct R&D investment is 9 percent or 2 percent. Without such basic knowledge it's almost impossible to manage R&D.

THE RQ SOLUTION

I confronted this measurement problem firsthand during a prior career managing missile guidance projects at Hughes Aircraft Company. Toward the end of my time there, I could see that changes in government acquisition policies were changing companies' incentives to conduct R&D. I could further see that Hughes's responses to these policies, as well as the company's response to being acquired by General Motors, were dramatically changing the way we organized R&D. I was concerned these changes would permanently degrade Hughes's R&D capability. Moreover, I suspected this was true not only for Hughes, but for all companies in the defense industry, and possibly other industries as well. The challenge at the time was that I couldn't convey the need for alarm. Without a good measure of R&D capability, there was no way to demonstrate there was a problem.

I became an academic in part to solve the R&D measurement problem. While the original insight for the solution occurred to me as a first-year PhD student, the hard work to implement and validate it has taken 20 years. The result of that work is a measure called RQ™ (short for research quotient)—a name intentionally similar to individual IQ, to reflect the fact that both measure problem-solving capability. In fact, I originally called the measure IQ (innovation quotient).[7] For individuals, IQ is measured as the speed and accuracy of solving problems of increasing difficulty. Within any given time constraint, individuals with a higher IQ solve more problems correctly than those with a lower IQ. For companies, RQ is efficiency solving new problems. For any given level of R&D spending, high RQ companies will generate more innovations, or for any given innovation, high RQ companies will invest less developing it. Accordingly, RQ is mapped onto the IQ scale (mean = 100, standard deviation = 15) to reinforce that intuition.

I argue that RQ is the most intuitive measure you could construct for R&D effectiveness. It captures a company's ability to generate value from its R&D investment in a very precise way. In particular, RQ is the percentage increase in revenue a company obtains from a 1 percent increase in R&D, while keeping everything else the same (the mathematical details are in Chapter 10).

Because RQ relates R&D to revenues, a company can have high RQ either by generating a large number of innovations and being reasonably effective exploiting them, or by generating a smaller number of innovations and being extremely effective exploiting them. One thing to note with this definition is that a company with a large number of patents or new products may not have a high RQ if it operates in small markets.

Accordingly, RQ doesn't fit everyone's definition of innovation. Some people prefer to think of innovation as the number of new things a company introduces. While this is important for some purposes, this isn't what RQ measures. Instead, RQ measures how much economic benefit the company derives from its R&D. The benefit *can* come from new products or services. However, it can also come from process innovation. In the case of product/service innovation, the economic benefit is reflected in higher revenue; in the case of process innovation, the economic benefit is reflected in lower costs.

What makes RQ so powerful as a measure of R&D is that it's derived from the "production function" in classic economics that relates a company's inputs to its output. This means that once a company knows its RQ, it can use economic relationships to forecast not only additional revenues from its R&D, but also profits, market value, and growth as well. Moreover, it can determine the optimal level of R&D investment. For all companies, there is a point at which the additional gross profit from R&D falls below the investment required to generate those gross profits. RQ allows companies to identify that optimal point precisely.

In addition to having a solid economic foundation, RQ solves the three problems with the patent measures. Because it is estimated entirely from standard financial data, RQ is *universal.* It can be computed for any company engaged in R&D. Second, because it is essentially a sophisticated ratio of output dollars to input dollars, RQ is unitless. Thus its interpretation is *uniform* across companies within an industry as well as across industries. Perhaps most important, RQ is a *reliable* measure of R&D productivity. The theoretical predictions relating RQ to company R&D investment, market value, and growth held up when tested with 47 years of financial data for all publicly traded companies in the United States.

Finally, RQ solves the main problems discussed earlier for the 250-plus measures currently in use: (1) the data to derive RQ are *easy to collect* (in fact, for public companies, it is data they are required to collect and report in their 10K), (2) there are *uniform standards* across business units that are imposed by FASB, (3) RQ can't be *gamed* to make sure a group looks good—the only way to look good is to have higher output or lower input costs, (4) RQ defines the relationship between inputs and outputs (rather than focusing on one or the other), and perhaps most important, (5) RQ is meaningful to shareholders, because now they, too, can predict how R&D spending will affect stock price. Moreover, as we will learn in Chapter 8, RQ can be used by investors to outperform the market.

Can't We Just Use Intuition?

In principle, measures aren't necessary if managers have good intuition about what bets to place and how to execute them. Certainly, none of the most famous entrepreneurial innovators needed measures of their R&D productivity: Henry Ford, Thomas Edison, or more recently Steve Jobs, Bill Gates, and Jeff Bezos. Each of them had a clear vision of what new product or service was needed, how valuable it would be, and what was necessary to execute that vision.

But what about the rest of us? How good are non-unicorns at gauging what drives innovation? Let's see. Test your own intuition by answering 12 questions about factors that are often associated with innovativeness (Figure 1-2).

Now check how you did. Give yourself one point for each answer that matches the answers in Table 1.1, column C. Column C provides the "truth"—what the data tell us about factors enhancing innovation. I will provide details on how I arrived at the truth in the remaining chapters.

FIGURE 1-2. **Survey of R&D practice effectiveness**

RQ is a new measure of firms' R&D productivity that reliably predicts R&D spending, profits, market value and growth. While we have the RQs for all US-traded firms. We are trying to understand what factors drive RQ, and would like your help.

1. To what extent do you think each of the following factors affects RQ

	RQ significantly decreases with factor	RQ decreases with factor	No effect	RQ increases with factor	RQ significantly increases with factor
Age (years since founding)	○	○	○	○	○
Firm Size (revenues)	○	○	○	○	○
Competition (number of firms in industry)	○	○	○	○	○
CEO compensation (% tied to stock price)	○	○	○	○	○
R&D horizon (% spent on research rather than development)	○	○	○	○	○

2. CEO BACKGROUND. Which type CEO is associated with higher RQ

○ Internal CEO

○ Hired from another firm

3. R&D DECISIONS. Which method of allocating R&D is associated with higher RQ

○ R&D decisions centralized

○ R&D decisions made by divisions

4. R&D LOCATION. Which form of R&D is associated with higher RQ

○ Internal R&D

○ Outsourced R&D

5. R&D RISKINESS. Which type of R&D is associated with higher RQ

○ Radical Innovation (new to the market)

○ Incremental Innovation

6. R&D SPENDING PROFILE. Which R&D spending profile is associated with higher RQ

○ Fairly consistent percentage of revenues each year

○ R&D budget varies with opportunities

FIGURE 1-2. Survey of R&D practice effectiveness, *continued*

8. OPERATIONS. Which top-level operational strategy is associated with higher RQ

- ◯ Operations domestic to the US (other than exports and outsourcing)
- ◯ Firm has operations subsidiaries in multiple countries

9. Please tell us about your current role

- ◯ Firm management (non-R&D)
- ◯ R&D management (central R&D function)
- ◯ R&D management (division or project level)
- ◯ Investment management
- ◯ Investment analyst
- ◯ Government policymaker
- ◯ Consultant
- ◯ Academic

Other (please specify)

TABLE 1-1. Answers to Survey of R&D Practice Effectiveness

	MANAGERS (NON-R&D)	INVESTMENT COMMUNITY	THE DATA
Company age	2.9	3.2	1
Company size	3.4	3.1	5
Number of competitors	3.8	3.0	5
% of CEO pay tied to stock	2.8	2.1	3
% of R&D that is R (vs D)	3.3	3.5	5 (up to a point)
Internal CEO	33%	90%	Internal CEO
Centralized decisions	47%	10%	Centralized
Internal R&D	61%	89%	Internal R&D
Radical Innovation	50%	30%	Incremental
Consistent R&D	50%	40%	Consistent
Score (closer to truth)	**6**	**4**	

Be generous—mark either 4 or 5 correct if the correct answer is 5, and mark either 1 or 2 correct if the correct answer is 1. If you scored a 9 or 10 you have exceptional intuition and may not need this book other than to provide deeper understanding of why your intuition is correct. If you scored 7 to 8, your intuition is better than the average manager's, but you can still benefit from understanding where your intuition is misleading you. If you scored 6 or below, you fall into the majority of managers and financial professionals who need a reliable measure to help make R&D decisions and gauge their effectiveness.

Now let's dig deeper. Rather than looking at overall score, look at your answers to specific questions. If you're like most managers inside operating companies, you provided the answers in column A. If you're like most professionals in the investment community, you provided the answers in column B. So the first thing to notice from comparing the two columns is that managers and investment professionals have different views on what makes R&D productive. For example, managers believe competition leads to greater innovation, while finance professionals believe it has no effect. Managers believe *outside* CEOs drive greater innovation, while investment professionals believe *internal* CEOs drive greater innovation. Perhaps most notably, investment professionals strongly believe *decentralized* decisions lead to greater innovation, while managers are split on whether centralized or decentralized R&D is more effective. These differences between the two sets of intuition could lead to problems if companies want to go left, but investors pressure them to go right.

The second thing to notice is that managers have only slightly better intuition than investment professionals: managers better gauge the truth for 6 of the 10 factors, while investment professionals better gauge the truth for 4 of the 10 factors. That should come as a surprise. Managers operating inside companies are much closer to "the innovation phenomenon" than investors, so they have more opportunity to develop and refine their intuition. The fact that managers don't have much better intuition suggests that being closer to the phenomenon doesn't help much. There are very few domains in which expertise and experience are unhelpful. The most notable popular example comes from Michael Lewis's book *Moneyball*,[8] where experienced baseball scouts were poorer at choosing players than the sabermetrics employed by Billy Beane. Not surprisingly, the solution to poor R&D intuition is the same as the solution in *Moneyball*—replacing intuition with data and meaningful measures.

The third and most important point is that neither managers nor investors have great intuition. This explains why in the absence of a measure like RQ, it was easy for R&D productivity to decline so dramatically. While companies implemented practices they believed would improve innovation, they had no way to gauge whether those practices were making their innovation more or less effective.

A final problem is that even for the factors where R&D managers' intuition is solid, without data and measures it's difficult to defend against investor pressure to comply with factors where the intuition of the two groups differs.

This first test of innovative intuition is admittedly difficult, so I'm going to pose a second test. Rather than identifying specific practices that increase or decrease innovation, I'm going to ask you to identify innovative companies. The logic here is that it should be easier to identify a "black box" than to know what's inside. For this second test, rank order the following companies in terms of their R&D productivity.

3M
Amazon
Apple
Google
IBM
Microsoft
Tesla Motors
GE
The average company conducting R&D

Now score yourself on this test by comparing your rankings against those in Table 1-2, which shows the true ranking of these companies' innovativeness using their RQ.

TABLE 1-2. RQ Ranking of "Strategy& Innovation 1000"

	RQ RANK	STRATEGY& RANK
Amazon	22	4
Google	38	2
Apple	42	1
Microsoft	89	7
GE	185	6
Average R&D company	209	500
Tesla	213	9
3M	218	5
IBM	224	8

Give yourself a point for each company that you correctly identified as being above or below average (rather than worrying about the full ranking). If you scored five points you should feel pretty good about your intuition. But you shouldn't be too disappointed with a score below this. *All* these companies were ranked within the world's 10 most innovative companies by respondents to the 2015 Strategy& Global Innovation 1000 survey. As you can see, however, none of these companies are in the top 10 with respect to R&D productivity. Worse, three of them are actually below average. So even innovation experts have unreliable intuition.

Why is our intuition so poor at both recognizing innovative companies and identifying the factors that make them innovative? The answer to the second question is fairly easy. Unless we work inside a company, we know almost nothing about it other than what the company chooses to disclose. Given that we can't identify who's innovative, it would be almost impossible to identify what factors make them innovative. You can't isolate best practices until you actually know who exemplifies them.

It's fairly easy to understand why there is little hope for outsiders to gauge company innovativeness, but what about folks inside companies? We saw from Table 1 that managers don't have much better intuition than investment professionals. How is that possible? The most obvious explanation is that the consequences of R&D practices often aren't felt for a number of years. By then a number of other things have changed, both internally as well as in the market. Thus it is very difficult to identify how much an increase in sales is due to changes made three years ago to the R&D practices, versus to changes in marketing, in the number of competitors, in those competitors' products or services, or in consumer tastes.

Contrast this with advertising. All the impact of advertising is typically felt within a few months—a time so short that little else has had a chance to change. Accordingly, companies' intuition about how much to spend on advertising, how to allocate that expenditure, and how effective a given campaign was is fairly reliable. In fact, in the context of advertising, we can rely on "knowledge" rather than "intuition."

FIGURE 1-3. **HP's R&D intensity decreased steadily since John Young's tenure as CEO**

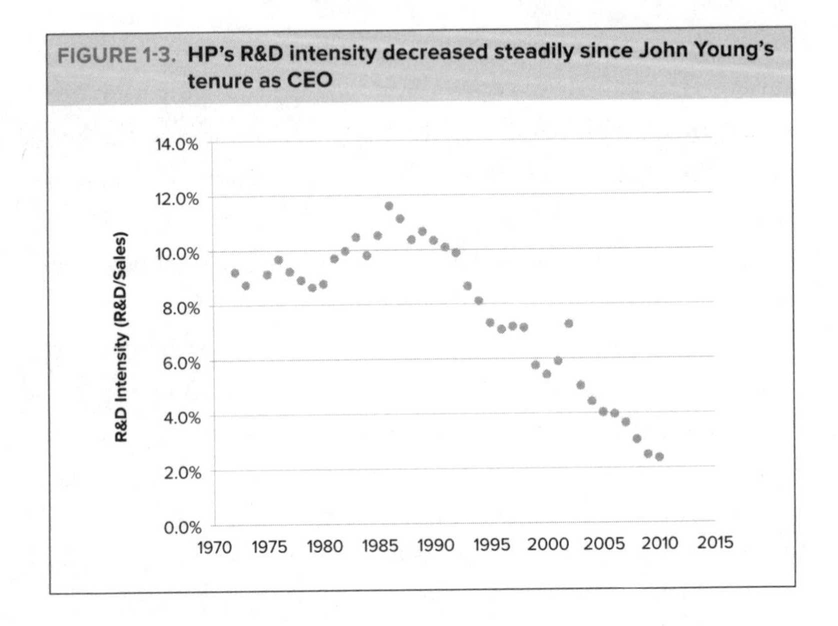

HP Epilogue

Now that we have the RQ measure and understand why we need it, let's return to the case of HP. Was Mr. House right? Did Mark Hurd destroy the innovation engine that made HP great? Let's look first at R&D spending (Figure 1-3). Certainly HP's R&D intensity (R&D/Sales) has decreased. Throughout the 1980s R&D spending averaged 10 percent of sales (even higher than Mr. House remembered), but by 2009 it had fallen to 2.9 percent of sales. This wasn't a Mark Hurd effect, however. Figure 1-2a shows the 2009 level merely tracked a steady decline that began in 1993. Thus Mr. Hurd was sustaining an R&D strategy implemented by Lew Platt (appointed CEO in 1992) and perpetuated by Carly Fiorina (appointed CEO in 1999). Note the decline was in R&D intensity, not in absolute R&D investment. R&D spending continued to grow, but

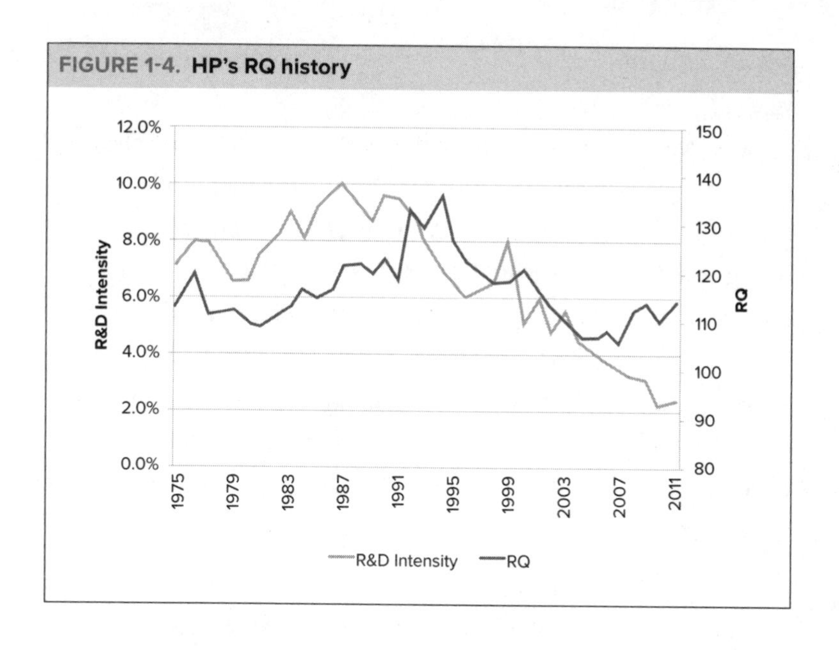

because the company undertook a "growth through acquisition" strategy, revenues grew more rapidly than R&D.

We defer to Chapter 4 the question of whether 9 percent or 2 percent of sales was the correct level of R&D investment and turn to the question of whether the innovation engine deteriorated. To evaluate that we examine the trend in HP's RQ over time. Figure 1-4 shows that HP's RQ reached a peak of 137 in 1994 (one of the highest RQs in the economy), so Mr. House is likely correct that R&D was one of the things that made HP great. It then followed a steady decline to 107 in 2005, the year Mark Hurd became CEO. However, it actually *increased* to 114 under his tenure!

This means HP's RQ went from being in the top 2 percent of companies to being in the top 33 percent (still very good, but certainly not as great). In Chapter 9, when we discuss how to change RQ, we will speculate what (besides the investment decline) might be responsible for the 30-point decline over 20 years.

So we see how RQ can be used retrospectively to diagnose problems. Of course that's all we can do at this point, because companies haven't had the measure until recently. For the sake of argument, let's assume that RQ was available 30 years ago. How might that have changed company behavior? I believe it's likely companies would have made very different strategic decisions—decisions that would not only have preserved their own innovative capability and growth, but as a by-product would have preserved economic growth. Let's speculate what might have happened differently at GE.

What If Jack Welch Had Had the RQ Measure?

GE was a U.S. innovation treasure for a good part of the twentieth century. Indeed, its genesis was Thomas Edison's lightbulb. It had one of the best central research labs in the country, reinvested 40 to 50 percent of profits in R&D each year, and had one of the highest returns on that investment in the entire economy—double the returns of the average company today.

However, that changed when Jack Welch adopted the market power strategy (be number one and number two in the market) advocated in Michael Porter's popular book *Competitive Strategy*. R&D plays no role in that strategy because it is a strategy about maximizing current profits. To execute its strategy, GE divested many businesses that had been reliant on R&D (televisions, semiconductors, and aerospace) and expanded into businesses where R&D had no role: television broadcasting (NBC) and finance (Kidder, Peabody).

So GE began cutting R&D (as a percentage of profits) from a high of 52 percent to below 15 percent. It divested Sarnoff lab when it acquired RCA, and it began outsourcing what R&D remained. In the end, GE had gutted one of the country's greatest innovation engines, to a point where its R&D productivity was one-third its peak.

There is no question shareholders enjoyed a tremendous ride. GE's stock price increased from $1.30 in April 1981 when Jack Welch took office to a peak of $58.00. However, one of the costs of that ride was GE's innovation engine, so by the time Jack Welch handed the baton to Jeffrey Immelt, the stock had fallen to $37.00. Worse, with no source of internal growth the stock continued to fall to a low of $10.00.

Had Jack Welch known GE's RQ in 1986 when he began reducing R&D, he might have taken a different course. He would have known that maintaining R&D would have provided even greater growth and market value. Had he further known GE's RQ in the 1990s, he might have recognized that outsourcing R&D was dramatically reducing its returns and might therefore have kept R&D in-house. Thus this inability to measure innovative capability may have contributed to GE's decline.

SUMMARY

HP and GE are just examples. While they offer some rich detail and accordingly clues as to how innovative capability declines, it is hard to generalize from them. Unfortunately, similar case studies are where most innovation prescriptions come from. This book takes a different approach. Rather than recommending insights from these cases as prescriptions for improving innovativeness, I treat these insights as hypotheses—educated guesses about what drives RQ. I obtained other hypotheses from my own experience in industry, from deep understanding of the economics and management literature on innovation, and from interviews with all levels of management in R&D organizations as part of an initial National Science Foundation (NSF) grant.[9] This grant allowed me to generate rich case studies of high RQ versus low RQ companies from which to identify factors driving company innovativeness.

Accordingly, the first NSF study was very similar to other case studies that form the bases for innovation prescriptions. However, as I mentioned, I didn't generate prescriptions from this study. I only generated hypotheses about what factors might be driving companies' innovativeness. To understand how broadly applicable these factors were, I applied for and obtained a second NSF grant to test these hypotheses across the full spectrum of U.S. companies engaged in R&D.

This study not only allowed me to identify which companies were most and least productive with their R&D investment, it also allowed me to determine what R&D practices they follow. This is ordinarily extremely difficult for outsiders, because these practices are held pretty close to the vest. However, through the grant, I had access to confidential data on companies' financial information as well as their R&D practices from the NSF Survey of Industrial R&D (SIRD) and the Business Innovation and R&D Survey (BRDIS). The financial data allowed me to characterize the RQ of all companies in the NSF datasets. Accordingly, I could identify companies that were highly productive as well as those that were less productive. By then matching the companies' RQs to their R&D practices, I could isolate what high RQ companies did differently from their low RQ counterparts that might drive the differences in their innovativeness.

As a result of these two NSF studies, I have been able to identify which popular innovation recommendations actually work and which don't. The results are startling. Many of the most popular recommendations *don't* work. Not only do they fail to improve innovation, in many cases they actually make companies worse at innovation! Thus experts may actually be contributing to the decline in company innovativeness.

Because of these NSF studies, *How Innovation Really Works* is unlike any other innovation book. Rather than offering prescriptions I *believe* are correct, I offer prescriptions I *know* to be correct on average, because I have tested them against the entire spectrum of U.S. companies doing R&D. The book combines the results from those studies with other foundational work to help companies design the comprehensive innovation systems they need.

In this sense, *How Innovation Really Works* is like sabermetrics for R&D companies. Just as TQM reversed the decline in companies' manufacturing productivity, the hope for RQ is that it can help reverse the decline in companies' R&D productivity. This book shows companies how. It provides actionable information to increase companies' market value in the short run through wiser R&D investment. In addition, it allows companies to benchmark their capability to see if there is opportunity for improvement in the long run. Furthermore, it tells companies how much improvement they can likely achieve, provides prescriptions on how to achieve that, and identifies how much that improvement should increase revenues, profits, and market value. While this prospect should be promising to managers and shareholders, it has even broader implications. Because R&D productivity drives economic growth, reversing the decline in company innovativeness has the potential to restore economic growth as well. If so, Robert Gordon's dismal claim in *The Rise and Fall of American Growth*, that the rapid growth of the twentieth century won't be repeated, may be proved wrong.

MISCONCEPTION 1: Small Companies Are More Innovative

Many people believe small entrepreneurial companies are the real source of innovation and economic growth in the economy. In fact, this is the main message in Gary Shapiro's book *Comebackː How Innovation Will Restore the American Dream*.[1] Ironically that view conflicts with that of the economist most often associated with entrepreneurship, Joseph Schumpeter. Schumpeter argued instead that large companies are the major engine of economic growth.[2]

THE COMPANY SIZE DEBATE

Large Company Advantage

Schumpeter's argument relied on two advantages of large size. The first advantage is scale economies. Since R&D is a fixed cost, the amortized cost of R&D per unit decreases as a company sells more units. Since companies with large manufacturing, marketing, and distribution systems are likely to sell more units, their R&D cost per unit is lower.

The second advantage of large size is that it is typically correlated with market share and pricing power. This means when a new product is introduced by a large company it is likely to command a substantial price premium. These two "Schumpeterian advantages" together mean large companies are likely to have both lower costs and higher prices per innovation than small companies.

While Schumpeter's arguments for large company advantage deal with ability to exploit innovations, there are also advantages to large size in generating innovations. First, there may be minimum efficient scales for some R&D projects. For example, the cost to build and run a supercollider is so prohibitive there is only one in the world. The Large Hadron Collider required 10 years and $4.75 billion to complete, and the estimated cost to operate it is $1 billion per year. While this scale is so large that not even a big company can afford it, there are other technologies large companies can afford, while small companies can't.

Second, R&D projects are known to involve a high degree of uncertainty. Stevens and Burley[3] report that on average, for Industrial Research Institute (IRI) member companies, it takes 125 funded projects to achieve one commercial success. Large companies deal with this uncertainty by creating large portfolios of projects. Companies diversify their R&D project risk in the same way investors diversify their financial risk—by holding a portfolio. Small companies, in contrast, can't pool their risk—all their eggs are typically in one project basket. Accordingly, their survival often depends on the outcome of that project. As an example, of the 1,800 biotechnology companies founded since 1980, fewer than 200 generate positive revenue and only 6 are net profitable.[4]

What's particularly appealing about the large company R&D portfolio is that while companies need it to diversify risk, once they've diversified in this fashion, they have another advantage: technical diversity. This technical diversity implies a broader set of problems and associated technical expertise,[5] which increases the likelihood of having any required expertise in-house.

Finally, large scale implies a broader set of product markets. Drawing again upon the uncertain nature of R&D, this increases the likelihood that projects that fail for a given application might have other applications elsewhere in the company. Take for example the case of ion beam thruster technology at Hughes Aircraft Company. The technology was originally developed for attitude correction in military satellites. It was ultimately deemed infeasible for that application because military satellites had a five-year life, which was insufficient to pay back the high cost of the technology. However, when engineers at the company's Santa Barbara Research Center (SBRC) learned of the ion beam technology, they recognized its value for implanting layers on semiconductors. The technology allowed them to create larger and higher-resolution infrared detectors, contributing to SBRC's designation as "the industry's premier detector facility."[6]

Small Company Advantage

On the flip side of the size debate are arguments suggesting small companies are more productive with their R&D. Indeed, colloquially, people use the term "entrepreneurial" (conveying small startup companies) to connote innovativeness. The small company advantages typically pertain to organization design. One important element of organization design is compensation systems. Small companies have a number of advantages over large companies in creating compensation systems that align employee behavior with company goals. First, because there are fewer employees, it is easier to observe employee behavior directly. Accordingly, it is easier to design compensation systems that are based on observed behavior, such as how many hours an employee is working productively. Second, often there is only one group or department in the company. This means company-level outcomes are a reasonable measure of performance for that group. In other settings compensation tied to company performance is less effective because that performance is influenced by unrelated groups.

The ability to have compensation tied more closely to employee performance typically means that a company can attract the highest caliber employees. This is because high-caliber employees earn more when being paid for performance than they do when working for salary alone. In other firms, salaries are necessarily pinned to the average employee. A second important benefit of tying compensation to performance is that because employees are paid for their performance, they work harder.

A clever study by Todd Zenger, author of *Beyond Competitive Advantage*,[7] tested this idea that small companies have compensation advantages relative to large companies. Todd looked at engineers who started out at one of two large Silicon Valley companies. He then looked to see who left, where they went, and how much they earned.[8] Doing this allowed him to test whether small companies paid more, and accordingly whether that allowed small companies to attract higher caliber employees than large companies. He found indeed that employees earned more when they left, and that the premium for leaving was twice as high if they went to small companies than if they went to medium or large companies (roughly $10,000 versus $5,000 in 1990 dollars). He also found that small companies were more likely to offer equity and other forms of compensation tied to individual and group performance. Thus, small companies could attract higher caliber employees. On average employees in small companies had higher undergraduate grade point averages, higher degrees, and more awards, publications, and patents. Finally, employees in small companies worked harder (more hours) than those in large companies.

An additional organizational advantage small companies enjoy is better communication. Having fewer employees implies decision makers are closer both to the technology and to the customer. This means they can better link technological possibilities to market needs. In addition to making better decisions, these decisions are made more rapidly because there are fewer levels of hierarchy. Beyond these "vertical advantages" that stem from few levels of hierarchy are horizontal communication advantages. In a classic study, *Managing the Flow of Technology*,[9] Tom Allen found that when engineers tackle technical problems, their success depends on the resources they consult. The likelihood that they consult other employees turns out to be highly dependent on how close they are physically. In small companies, everyone is closer physically, so the likelihood of consulting other colleagues is high. While this study was published in 1977, a more recent study by Chris Liu indicates this is *still* true even though researchers now have e-mail and the Internet. While proximity had no effect on e-mail communication among researchers in the same lab, e-mail communication between workers in different departments increased threefold when they were on the same floor, and fivefold if they were on the same wing of that floor.[10]

RESHAPING THE SIZE DEBATE

Thus there appear to be compelling arguments for large companies or small companies to be more effective in conducting R&D. This likely explains why we see both sizes engaged in R&D. Accordingly, more recent theories attempt to reconcile the coexistence of large company and small company R&D. These theories suggest the two sizes differ in the type of R&D they conduct.

One way that large companies differ from small companies is that they conduct more basic research.[11] *Basic research* is defined by the National Science Foundation (NSF) in its Business R&D and Innovation Survey (BRDIS) as the planned, systematic pursuit of new knowledge without specific immediate commercial application.[12] It is distinguished from *applied research* (planned, systematic pursuit of new knowledge aimed at solving a specific problem or meeting a specific commercial objective) and *development* (the systematic use of research and practical experience to produce new or significantly improved goods, services or processes).

The vast majority (82 percent) of R&D in companies is development. One example of basic research is the 1960 invention of the hydrogen maser, an atomic clock that's now at the heart of satellite-based global positioning systems (GPS). The notion that the invention precedes application is nicely captured by its inventor, Dan Kleppner: "I wasn't dreaming of developing the GPS. With basic research, you don't begin to recognize the applications until the discoveries are in hand."[13]

Because basic research may take 10 years or more to contribute to something commercially viable, it is most often done by universities and government labs. However, back in the 1920s, companies began to engage in basic research to create future opportunities. DuPont, for example, in 1926 established a program in fundamental research to "establish or discover new scientific facts without obvious practical applications."[14] Beyond its potential for generating commercial opportunities for DuPont, it was felt the program would enhance the company's prestige and facilitate the hiring of scientists. One of the first areas of basic research at DuPont was polymer science. That science ultimately led to the invention of neoprene in the early 1930s and nylon in the mid-1930s.

Large companies are more likely to conduct basic research because they have a broader technological base to help them identify where fundamental science prohibits further advance in their existing product lines. In addition, large companies have a wider range of product markets they can profitably pursue if the research identifies unforeseen opportunities.

The second way company size affects R&D is that it determines whether companies conduct incremental innovation rather than radical innovation. We devote an entire chapter to radical versus incremental innovation in Chapter 5. Here however we merely want to point out that large companies engage primarily in incremental innovation because the returns to incremental innovation increase in the scale of the product or service that is being improved. In addition, incremental innovation is less risky. The company already knows who the customers are, and it has systems in place to reach them. The company likely also knows what feature improvements the customer values most highly. Take, for example, Apple's introduction of the iPhone 6s. The company already had large-scale manufacturing of essentially the same form factor (the iPhone 6), and it had a comprehensive distribution system (the Apple stores and the service providers). It likely also knew how many existing customers would upgrade to the new phone, and how many customers of rival phones would likely switch. Compare this to the introduction of the iPod in 2001, where the only prior product had been a Mac computer distributed through electronics retailers. In contrast, precisely because they lack scale, small companies require riskier projects that have the potential for greater price premiums.

The final way company size affects R&D is that it determines whether companies conduct process innovation to reduce the cost of their existing products, or product innovation to generate new or improved products. In logic similar to what we discussed for incremental innovation, large companies prefer process innovation. Because process innovation reduces the cost of products sold to customers, the returns to that innovation increase in the number of customers. In contrast, returns to product innovation typically stem from sales to new customers, and thus are independent of the company's current scale. Because small companies have small current scale, they prefer product or service innovation because of its higher margins.

THE RECORD ON COMPANY SIZE

The company size theories are one of the most tested sets of theories in innovation economics. Despite numerous tests, the debate on the optimal company size for innovation had been unresolved until recently. Indeed, the stylized facts characterizing the findings from those tests posed a puzzle. Tests of innovative *behavior* indicated that R&D investment increased with company scale, while studies of innovative outcomes tended to find small companies were more productive.[15] Accordingly, we were left with the puzzle of seemingly irrational behavior of large companies: R&D investment increases with company scale, even though R&D productivity appears to decrease with scale.

Generally, when academics obtain a result suggesting that companies are irrational, there is something wrong with the tests. The most obvious candidate in this case was the measures (product counts or patent counts) used in studies showing small companies were more productive. Now that you know that large companies do more incremental and process innovation, you see why using patents and product counts might be problematic. Process innovation doesn't show up in the product counts, and companies typically don't patent either their incremental or process innovation.

When I retested the theories using RQ, I found that large companies are in fact rational. They both invest more in R&D *and* have higher RQ. On average, RQ increases with scale throughout the entire range of company size. To provide a sense of proportion, if we define large companies in the same way the Small Business Administration (SBA) does, companies with more than 500 employees, then large companies have 13 percent higher RQ than small companies.

Large companies are more innovative for all the reasons the theories suggest. They do more basic research, more process (rather than product) R&D, and more incremental (rather than radical) innovation. Each of those forms of innovation has higher returns than the forms preferred by small companies, and the returns to each form of innovation increase with the scale of the company.

INTERPRETING THE RECORD ON INNOVATION AND SIZE

Given that large companies have more productive R&D, why are small companies considered to be more innovative?

Small Companies' Innovation Is Easier to See

For one thing, small company innovation is often more visible. Small companies have to swing for the fences to attract any market share from large companies, and home runs attract attention. The problem with swinging for the fences, however, is that the probability of hitting a home run is extremely low—2.67 percent for the 2015 MLB regular season. Similarly, the probability of small company success is low—on average only 25 percent of venture capital (VC) backed startups ever return their invested capital. Further, VC backed companies are a very select group—fewer than 5 percent of companies receive venture capital. So while hitting over the fences attracts a lot of attention—it's still highly unlikely small companies can do it.

Small Companies Invest Less in R&D

A second reason small companies are considered more innovative is that their innovations seem to magically appear—in other words, they generate innovations without investing much in R&D. This makes their R&D look more productive than that of large companies. How do they get away with this? The truth is small companies "borrow" innovations developed by large companies. It is well known that Bill Gates didn't develop DOS, he bought it for $50,000 from Seattle Computer Co. Similarly, Steve Jobs imitated the Graphical User Interface (GUI) after seeing it at Xerox PARC and hiring employees who had formerly worked at PARC. Thus the innovations small companies are credited with often come from large companies. So even small company innovations are really large company innovations!

This leads to the reason large companies have higher RQ than small companies—they have "complementary assets" that help them better exploit the innovation. These include established supplier relationships, large-scale manufacturing, and well-developed sales, marketing, and distribution systems to both bring the product to the attention of customers and get it in the hands of those customers. To put these advantages in context, Merck's technical and marketing systems are so sophisticated that according to estimates in Hart Posen's dissertation, a blockbuster drug in Merck's hands is worth on average $2 billion, while in the hands of a small company, the same drug would only be worth $125 million, one-eighth the amount to Merck.[16]

We May Be Confusing Size with Age

A final reason people may think small companies are more innovative is that when they say "small" they typically are thinking "young." The general impression that small companies are more innovative stems from the tremendous success of a handful of startup companies: Amazon, Google, Apple (note of course none of these companies is small now). So one reason we believe small is better is that we equate small with startup. I think most people would believe that small companies that are older are less innovative—otherwise they would have grown larger.

Let's shift the discussion to young companies. One reason we believe young companies are more innovative is that we only see successful young companies. Unsuccessful young companies die quickly. As just mentioned, less than 5 percent of startups obtain VC financing, and of those that do, only 25 percent return their invested capital. So when we think of startups, we are likely ignoring 98.75 percent of companies. Few people see those 98.75 percent, and even those that do likely forget about them. The tendency is to see the companies that survive and assume they represent all startups. They don't! They're the exception.

Given the high odds of failure, it's clear that starting a new venture is extremely risky. Some companies are tremendously successful, and some are abysmal failures. Accordingly, young companies and small companies both have the highest range of RQs. Even among companies that are publicly traded (a tremendous badge of success in itself), the highest RQ and the lowest RQ both belong to small companies. As companies get older and larger, the range of RQ narrows, as shown in Figure 2-1. The distinction between size and age, however, is that while the range of RQs narrows with both size and age, on average RQ *increases* with size, but *decreases* with age. Across all public companies, an additional year of age decreases RQ 0.24 percent, on average. Thus companies maintain the advantage of youth while gaining the advantages of scale if they grow big fast, like Amazon, Google, and Apple.

One of the advantages of youth that Amazon and Google share is that they still have their founders. While we don't know all the reasons this is important, founders have stronger strategies (which is why they became successful) and better intuition about how all the pieces work and fit together (because they've managed them from the seed stage). Both of these lead to better management. In addition, however, the company's success allows the founders to maintain their strategies in the face of investor pressure for current profits.

Startup Innovations Don't Really Come from the Startup

But even if we narrow our discussion to successful startups, they appear more innovative than they really are because their initial innovation typically comes from outside the startup. We mentioned that Bill Gates purchased DOS and Steve Jobs imitated

FIGURE 2-1. **While RQ increases with company size (a), it decreases with company age (b)**

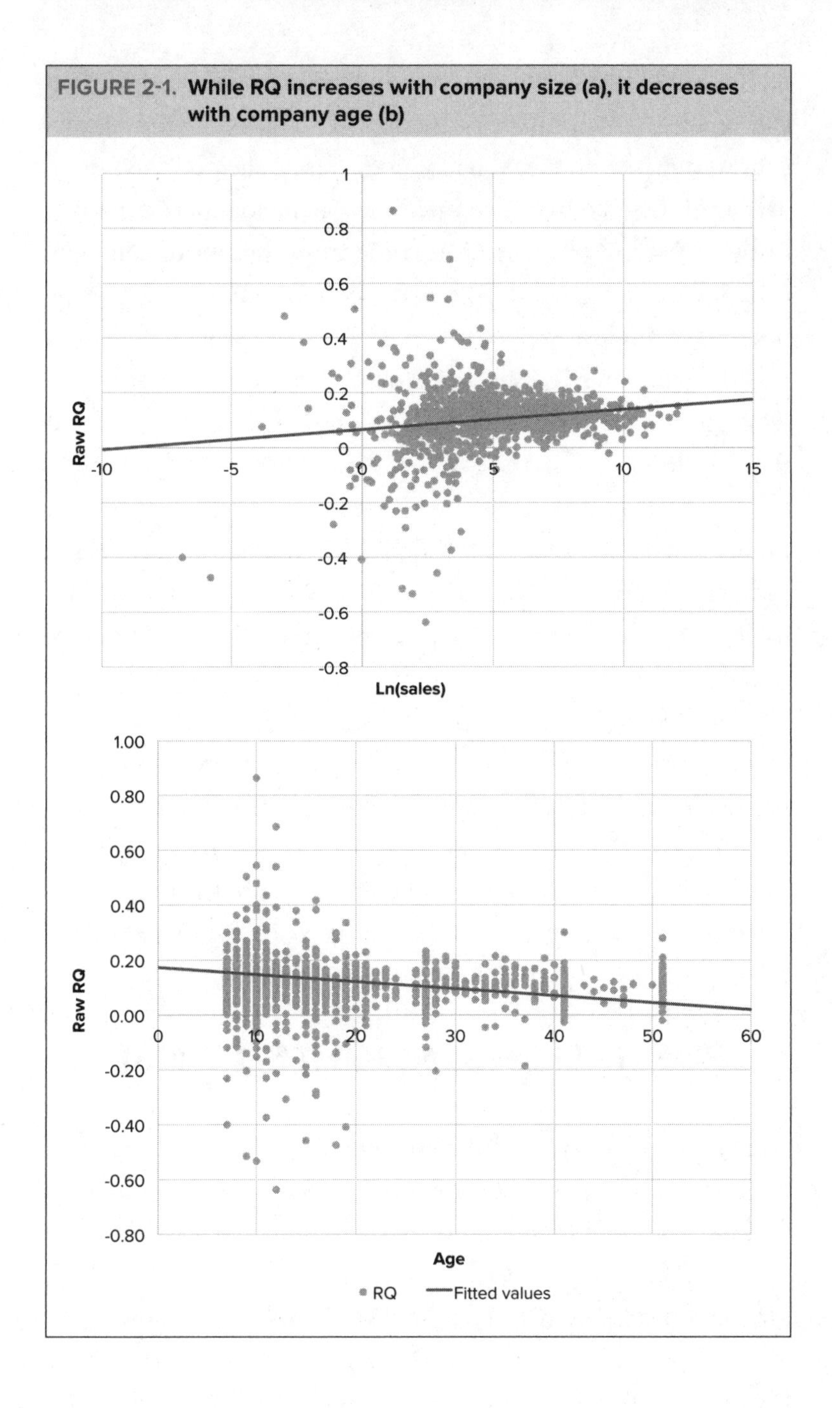

technology from Xerox PARC. They are the rule rather than the exception. In a survey of Inc. 500 companies,[17] only 21 percent were based on ideas the founders researched on their own. The most likely source of venture ideas is the founder's prior job. Fully 52 percent of those ideas come from their prior company or industry, and another 14 percent come from a buyer or supplier to that industry.

The late Steve Klepper conducted a series of industry studies examining how the founders' origins affected the likelihood they would succeed in their new ventures. What he found across these industries is that the most successful startups came from founders who had worked for successful companies in the same industry. What these founders brought with them was deep understanding of how the industry operates. They also brought with them a sense of how the product offered by the prior employer could be improved (auto startups),[18] as well as a sense of where there was opportunity for new products that weren't profitable for the former employer to pursue (laser startups).[19]

The point is that in most cases the innovativeness attributed to the startup venture actually should be attributed to an existing company.

PUTTING THE INSIGHTS TO USE

Now that we understand the advantages of each class of company size, let's see if we can incorporate those insights into making each size more innovative. While size may seem to be something companies are stuck with in the short run, there are means to enjoy many of the benefits of large size without actually growing. Similarly, while overall large companies have higher RQ, this is principally due to their ability to generate and exploit innovations. Small companies seem to have an advantage with respect to developing innovations (converting the inventions generated by large companies into commercially viable products and services). Thus there are also means to enjoy many of the benefits of small size without breaking up the company.

Making Large Companies More Innovative

Let's recap what large companies are particularly good at, and where they might need help. Their main advantages come from scale. They have more technology and enjoy substantial purchasing discounts, more efficient production, larger sales forces, more efficient distribution systems, and greater seller power. As noted previously, these advantages are so substantial that a drug in Merck's hands is worth eight times as much as the same drug in the hands of a company at the twenty-fifth percentile in revenues.

But even with all these advantages, there may still be opportunity to benefit from behaving more like a small company. One obvious candidate is increasing the odds of commercial success of R&D projects from 1:125 to maybe even 2:125. To understand how that's possible, we're going to visit Xerox. In 1988, in response to the book *Fumbling the Future*,[20] Xerox created an internal venture fund called Xerox Technology Ventures (XTV). The goal of XTV was different from other internal funds in that it wasn't soliciting new ideas; rather, it was "recycling" ideas that had been abandoned by PARC. In creating XTV, Xerox mimicked many of the features of traditional VC funds. It committed $30 million to a fund with a 10-year life. It hired outside VCs and compensated them with high-powered incentives in the form of "carried interest." What this means is that the general partners receive 20 percent of the equity gain if a venture is acquired or goes public. The XTV partners identified abandoned PARC projects they thought held promise. Employees who had worked on the projects previously could continue to work in their old lab to develop them but were employed and paid by XTV. However, like the partners, employees gained an equity stake in the venture. Once the venture reached scale, employees would transition to XTV facilities full-time. With the exception of employees continuing to work in their existing lab, these features of XTV are quite similar to traditional VC.

In other regards XTV benefited from features of large companies. Xerox allowed XTV to use its supplier base and purchasing systems, its manufacturing capability, and its sales and distribution systems. In addition, XTV employees continued to work in the company's labs. However, XTV ventures were unencumbered by much of the Xerox red tape—such as producing manuals for new products in 38 different languages.

The most notable project example in the *Harvard Business Review* case of XTV[21] is the Advanced Work Station (AWS) Project. This was a project Xerox originally abandoned because it was forecast to take an additional 36 months and require an additional $25 million investment. With this cash flow profile, AWS didn't cross Xerox's investment threshold.

XTV adopted the AWS project and was able to complete it within 18 months at a cost of only $4 million. At completion, Xerox acquired the project from XTV for $15 million. This was a win-win for everyone involved. The XTV employees split $2.85 million; the three partners split $1.63 million, and Xerox got a $25 million project for $8.5 million. While the Advanced Work Station project is a specific example, it was typical of XTV's success. When the fund closed in 1996, Xerox had earned an annual return of 56 percent—four times that of the average VC fund.

The structure and outcomes for XTV share features with another innovative large company from a non-R&D context—Schibsted, the Swedish news organization. Schibsted was one of the few news companies that not only survived the shift to digital news, but actually thrived on it.[22] In both Xerox and Schibsted there was a segregation of the innovative unit from the operating units. The function of the separation seems to be unbridling the innovative unit from standard operating procedures. These units were not only physically separated, they had authority to make decisions without approval from the larger organization. Finally, in both cases, the innovative unit could utilize invaluable assets of the larger organization, such as brand, purchasing, manufacturing, and sales. Thus both organizations enjoyed many benefits of small organizations, while also enjoying the benefits of established organizations—the best of both worlds.

Both XTV and Schibsted are examples of "Skunk Works." Skunk Works was the name adopted for Lockheed's Advanced Development Programs (ADP). ADP was created in 1943 to build what was to become the XP-80 fighter jet. The Air Tactical Service Command (ATSC) contracted Lockheed to develop a fighter that exploited the most powerful jet engine at the time, the British Goblin. AWP designed and built the plane in 143 days! (For comparison the F-35 took 10 years (1996–2006) from contract to rollout).

The phrase Skunk Works (which Lockheed trademarked), has since been used to refer to projects characterized by (a) a "viciously small" highly select group, (b) with high level (c-suite) access, (c) substantial decision authority, (d) lack of encumbrance by normal procedures, and (e) physical separation from regular operations. Occasionally people confuse Skunk Works with central labs, which also share the characteristics of a small highly select group, high level access, and physical separation. What distinguishes central labs from Skunk Works is that the labs are generally tasked with basic and applied research to generate inventions (often with 5- to 10-year horizons), while Skunk Works typically refers to development of existing inventions (often expedited).

Making Small Companies More Innovative

Now that we understand how to introduce elements of small organizations into large organizations to make them more innovative, let's examine how small companies can access the benefits of large organizations.

The first thing small organizations can exploit is the knowledge of large organizations. The broad term for this is "spillovers" because it is knowledge the inventing company doesn't intend other companies to access but has trouble protecting. We've already discussed one form of spillovers—employees leaving their prior employer and taking ideas they contributed to. This "employee carrier" model of spillovers also occurs when the small company hires employees away from the large company, as Steve Jobs did with Xerox PARC employees.

Another form of spillovers is reading the science and technology literature, both journal articles and patents themselves. While companies can't infringe on the patent, the disclosure document generally includes foundational information that isn't protected by claims. Relatedly, but not technically a spillover (since it involves a contract), small companies can scour the patents available for licensing from university technology transfer offices.

Another common means to access the benefits of large companies is to form joint ventures with them. This is particularly common between biotech companies who have promising drug compounds but limited capacity to exploit them and large pharmaceutical companies with well-honed systems to exploit drugs but withering drug pipelines.

Finally, a close analog to the XTV solution for increasing innovation in large companies is for small companies to seek corporate venture capital rather than independent venture capital. Corporate venture capital operates almost identically to independent venture capital, but the fund is owned by a corporation rather than a partnership. Startups often avoid CVC investment because they fear the large company may constrain their development as well as their exit options. But there are tremendous advantages to being part of the right CVC. As with XTV projects, the new venture typically gains access to the funding company's manufacturing, purchasing, distribution, and sales capability. That may now seem obvious given the XTV example. What is less obvious is that CVC funds are actually more successful than independent VC! Their returns on investment are higher, and what's most surprising is that the rate at which they take portfolio companies public is higher (35.1 percent for CVC versus 30.6 percent for independent VC)![23] This almost seems implausible because the venture fund is a sideline for the large corporation, while it is the bread and butter of independent VC.

Certainly one advantage ventures in a CVC enjoy over those in a VC portfolio is the ability to share the large company's resources, as I've just mentioned. This is a source of added value that the VC can never provide. However, a second advantage is that the CVC has deep knowledge of its industry because it is operating in it. While VCs also specialize by industry, because they aren't operating they are unlikely to ever achieve the depth of knowledge and connections of the large company. The deep knowledge can manifest itself as better ability to select portfolio companies, better ability to enhance their technology, better ability to take the technology to market, or all three.

SUMMARY

We've seen that contrary to popular conception, large companies are more innovative than small companies. This is principally because they have scale and scope economies that allow them to better generate and exploit innovations. These advantages cause large companies to favor incremental and process innovations because these forms of innovation benefit more directly from the scale advantages.

We also better understand the source of the misconception that small companies are more innovative. First, people typically confuse small companies with young companies, and indeed companies are more innovative when they are younger. Second, we only see the successful young/small companies and overlook the other 98.75 percent of startups that fail to receive VC funding and to return at least the invested capital.

We also understand that even though these small companies are the exception, their success does suggest strategies for how large companies can be more innovative. Similarly, we've seen strategies in which small companies can access the advantages of large companies.

One final point worth emphasizing is that when all this is taken into account, large companies are the chief engine of innovation! Not only do large companies (using the U.S. Small Business Association definition of more than 500 employees) conduct 5.75 more R&D in aggregate than small companies, they have 13 percent higher productivity with that R&D. Moreover, this merely captures the private returns to their R&D. A further benefit of large company R&D is that it generates the spillovers upon which small company innovation free-rides. Thus it appears Schumpeter was right: large companies are the major engine of economic growth.

MISCONCEPTION 2: Uncontested Markets Are Good for Innovation

The first session in my Corporate Strategy class each year is an introduction explaining the distinction between it and Competitive Strategy. Competitive strategy answers the question *how to compete* in a market, whereas corporate strategy deals with *which markets* to compete in. To start the discussion, I ask students what type of markets Michael Porter would recommend. This gives them a chance to recite the Five Forces from his book *Competitive Strategy*: high entry barriers, no close substitutes, weak buyers, weak suppliers, and few rivals.[1] In short, you want an industry that insulates you from any price or cost pressure. This is essentially the prescription in the more recent book *Blue Ocean Strategy*.[2] The popularity of *Blue Ocean Strategy* attests to the power and durability of Porter's basic insights.

I then ask the students how many of them have read Porter's later book *The Competitive Advantage of Nations*.[3] Typically only a few of them have, so I explain that the book documents

Porter's effort to understand why companies from particular countries come to dominate the world stage in a given industry. As examples of what Porter has in mind, answer the following questions. If you want a luxury watch, what country do you want it to come from? If you want a new automobile, what country do you want it to come from? The most common answers are Switzerland for the watch and Germany for the car. Thus students intuitively grasp the question Porter is asking: What makes Switzerland and Germany such great breeding grounds for watch and automobile companies, respectively?

To investigate this question, Porter and a team of students conducted case studies in 500 country-industry pairs. The result of that heroic effort was the "four diamond" framework of national advantage. The four diamonds are: *factor conditions; demand conditions; company strategy, structure, and rivalry; and related and supporting industries. Factor conditions* pertain to resources that are abundant within the country. In old trade theory, these would have meant natural resources such as lumber or oil that the country could export. In Porter's framework, these represent resources requiring national investment, such as transportation and communications systems.

Note that the other three diamonds in Porter's new framework map onto forces in his old framework: *demand conditions* map onto *buyers*; *company strategy, structure, and rivalry* maps onto *rivals*; and *related and supporting industries* map onto *suppliers*. What's really exciting about comparing these three diamonds to the corresponding three forces, however, is that the prescriptions for them are polar opposites across the two frameworks. In the five forces framework, you want weak rivals, suppliers, and buyers. In the four diamond framework, you want vigorous rivalry, innovating suppliers, and demanding buyers, because two of the factors (demand and rivals) force you to innovate, while the third (suppliers) generates some of that innovation. In short, new Porter turns old Porter on his head!

So now I'll pose to you the question, "You're now the CEO. You get to pick which markets to enter. Do you want to enter 'old Porter' markets or 'new Porter' markets? Which do you think generates higher performance?"

This turns out to be a trick question, because it depends upon what your goal is. If your primary goal is profits, then you want to be in an "old Porter" market. *Profits* increase as markets become more monopolistic, from an average return on sales of 11 percent for companies in the bottom 20 percent of industry concentration to 13.6 percent for companies in the top 20 percent of industry concentration (plotted as squares in Figure 3-1).[4]

However, if your primary goal is stock market returns, you prefer to be in a "new Porter" market. *Monthly stock returns*

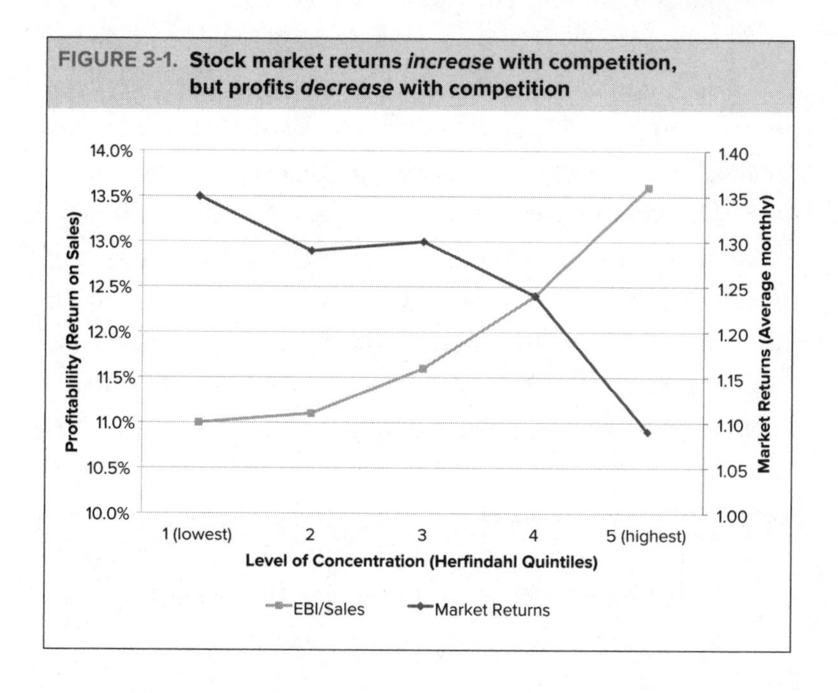

FIGURE 3-1. **Stock market returns *increase* with competition, but profits *decrease* with competition**

decrease as industries become more concentrated (less competitive), from 1.35 percent for companies in the lowest 20 percent of concentration to 1.09 percent for companies in the top 20 percent of concentration (as shown by the diamonds in Figure 3-1).

Why are the differences in financial performance so dramatic between these two types of industries? The answer is that the new Porter strategy focuses on growth—putting companies in an environment where if they don't grow they'll be eaten alive, while the old Porter strategy focuses on profitability—insulating companies from predators that could eat them alive.

SUCCESS AS THE SEEDS OF FAILURE

At this point you've likely thought of monopolists (and near-monopolists) that have grown dramatically: Standard Oil, AT&T, and US Steel in the early twentieth century, as well as Microsoft, Amazon, and Google in the late twentieth century. So you may be thinking that old Porter is still the way to go. There is even a view within economics that supports your intuition that innovation and growth are highest under a monopoly. This view was espoused almost a century ago by the economist Joseph Schumpeter,[5] to whom you were introduced in the last chapter. Ironically, Schumpeter is most often associated with the benefits of creative destruction and entrepreneurship (the opposite of monopoly). Schumpeter's argument was that monopolists not only benefit from the scale economies discussed in the last chapter, but they also benefit from the fact that without rivals, monopolists fully capture all the returns from their R&D. This is because they both serve the entire market and charge monopoly price. Under competition, rivals capture part of the market and drive price competition. Since monopolists have the highest returns to innovation, Schumpeter argues they should have the highest incentives to innovate. Combining the two theories suggests monopoly is a slam dunk. Not only does monopoly offer the highest profitability for the company (Porter), it also offers the highest innovation in the economy (Schumpeter).

But being a monopolist is a dangerous long-run strategy. For reasons I'll lay out in a minute, ultimately monopolists and near-monopolists stop innovating and accordingly stop growing. Then the real danger sets in. They become displaced by companies who enter from outside the market. The very insulation that sustains their profitability is the Achilles heel that leads to their disruption.

The theory that helps us understand why monopolists stop growing comes to us from Ron Adner, of *The Wide Lens* fame, and his former advisor, Dan Levinthal.[6] The theory had a different goal than understanding monopoly profits and growth. Its goal was to understand the forces driving a phenomenon called "technology curves." Technology curves are the common pattern we see for new technologies: their price typically decreases steadily after introduction, while at the same time their quality steadily increases. I'll refer you to the paper for details, but Ron and Dan's theory can generate these basic patterns of technology curves from just a few simple assumptions: (1) buyers differ in how functional a product needs to be before they are willing to buy it (as an example, most people didn't buy personal computers until they had a word processor and could substitute for typewriters), (2) buyers differ in how much they are willing to pay for that minimum level of functionality (some people bought personal computers when they were $5,000, while others waited until they could purchase them for less than $1,000), and (3) buyers differ in how much more they are willing to pay for the product as its functionality increases (for example, they upgrade when the processor is four times as fast, or when the newer personal computers can also support graphics).

To use a more current example, very few households own 3D printers at this point, because the consumer versions are both expensive and fairly crude (plug and play models cost $1,000 to $2,000). However, once their price drops, they become more user-friendly, and/or the number of applications increases, the market for 3D printers should expand.

So with these assumptions about how customers behave, companies in Ron and Dan's model decide each period whether they should conduct product innovation to increase functionality, or process innovation to decrease cost. As they make the decision, they try to maximize profits, taking into account (1) how many customers each strategy will bring into the market, (2) what strategy their competitor will follow, and (3) how many customers will choose each company's product.

The main result that Dan and Ron obtain from the model is that when companies face competition, they continually engage in innovation, because they not only try to bring customers into the market, they try to capture share away from rivals. This generates the "technology curves" in Figure 3-2a. When companies are monopolists, however, at some point they cease innovating because the value of bringing new customers into the market is lower than the cost of the innovation to accomplish that, as shown in Figure 3-2b. This is because these holdout customers have either extremely high functional requirements and/or require very low prices to entice them to buy the product. Once monopolists cross that threshold where the cost to innovate exceeds the profits from the additional customers they could capture, companies *rationally* stop innovating.

FIGURE 3-2. **Innovation continues under competition but ceases under monopoly**

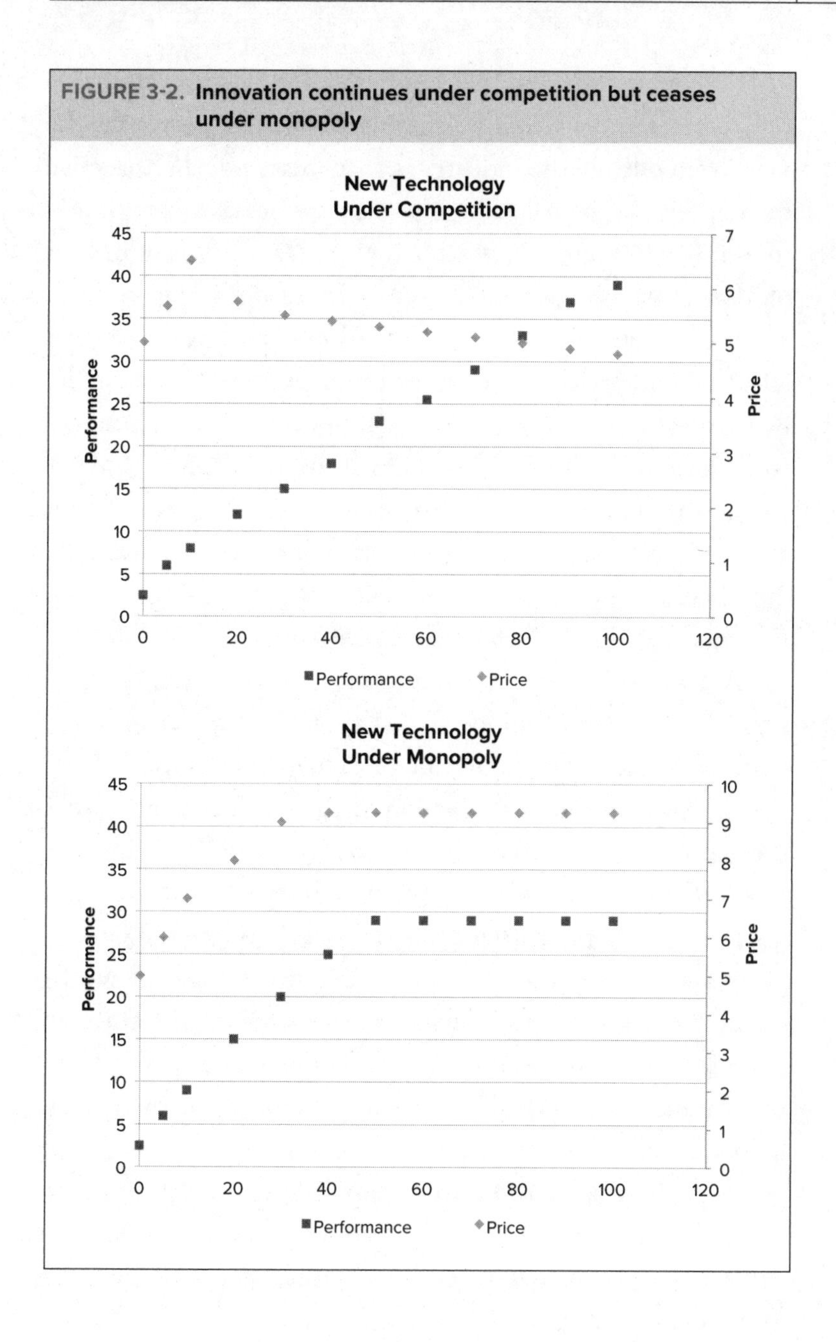

Of course the view that ceasing to innovate is rational is only correct if the world isn't going to change. However, the view from outside the industry is that this is a *really* attractive market—it's large, it has high profits, *and* it has a complacent (non-innovating) monopolist. It is precisely the type of market old Porter would advocate entering! It's also the type of market Richard Branson loves to enter. These are markets where "the customer has been ripped off or under-served, where there is confusion and/or where the competition is complacent,"[7] so Branson's Virgin Group comes in to play Robin Hood—stealing profits from rich monopolists by giving to the poor customers products or services at higher quality and lower cost. Thus because there are companies like Virgin, ceasing to innovate makes it highly likely the world *will* change.

In fact, these are the markets where we are most likely to see "disruptive innovation." What typically happens in these markets is that some group (niche) is being neglected because the monopolist chooses a version of the product that satisfies the greatest number of customers. Because disruptive innovations often serve such a niche, rather than the entire market, it is feasible for entrepreneurial companies to introduce these innovations at fairly low cost. Moreover, targeting a niche allows the entrants to circumvent the monopolist's entry barrier through some form of "end-run innovation." What makes this end-run innovation so formidable to the monopolist (and the reason monopolists are often wrongfully accused of being caught off guard by the innovation) is that adopting the entrant's end-run innovation requires duplicating existing investments, which by definition has negative returns. Thus the end run by the entrants forms an *adaptation barrier* to incumbents.

Netflix is a great example of a disruptive innovation. Its initial niche was people who found Blockbuster inconvenient, both because they had to drive there twice within a 24-hour period and because they had to scan a good portion of the stacks to find movies of interest. Formerly Blockbuster had an important resource (the store network) that formed an *entry barrier* to new companies. Any company considering a national chain would find the investment in stores unattractive—it would at best get half the market at lower margins (because the two chains would compete on price). So Netflix changed the rules of the game by executing an *end run* around Blockbuster's entry barrier through an online sales and mail distribution system.

Just as Blockbuster's store network served as an entry barrier, Netflix's end run became an *adaptation barrier* for Blockbuster. Because Blockbuster could already reach the entire market through its storefronts, the returns to investment in mail distribution were negative. The mail-order system wouldn't increase sales—it would *cannibalize* existing sales. Moreover, it might decrease sales of impulse purchases like popcorn or candy.

So let's think of other monopolists (or near-monopolists) who have been disrupted: the disk drive manufacturers in Clay Christensen's work[8] were disrupted by the smaller disk drives; major steel companies (which held a joint monopoly in the United States) were displaced by mini-mills; and Kodak was displaced when digital cameras were introduced. To be fair, none of these companies completely stopped innovating. However, they never seriously introduced the new technologies they invented because the additional market they could gain by introducing the technology wasn't sufficient to justify the investment (because they already had the entire market). In contrast, entrants had no market to lose by introducing the new technology, so those same investments had high returns for them.

HOW THIS RELATES TO RQ

Until this point in the chapter, you haven't heard a single word about RQ. However, the intuition I've just provided is about the returns to innovation, and since that's precisely what RQ measures, we should see that intuition show up in RQ. But before looking at how much competition affects RQ, let's first see how it affects R&D spending.

When I examined 50 years of data on publicly traded companies, I found that R&D spending increases as the number of companies in an industry increases. Thus Schumpeter was wrong about monopolists having the highest incentives to innovate. In fact, you probably knew intuitively that competition would increase innovation, even before seeing Dan and Ron's model. In general competition makes people and companies work harder.

However, one thing you might wonder about from Ron and Dan's model is whether the higher R&D spending that arises from competition generates any benefits for the company. All the companies are clearly spending more, but it seems plausible that spending more merely generates innovations that maintain the companies' respective market shares and profits. In essence, companies are playing a "Red Queen" game where "the hurrier they go, the behinder they get."

After having examined the 50 years of data, I now have an answer to that question as well. Competition increases not only the *amount* of R&D investment, it also increases the *returns* to that investment, RQ. This is a really satisfying result because it means that competition isn't ruinous. Instead, the increased effort people and companies expend under competition has a higher payoff than the effort they expend without it. This is precisely the result that Porter found in his *Competitive Advantage of Nations*. What the result seems to suggest is that while the competition may carve up the pie in smaller pieces, as it does so, it also grows the size of that pie. Higher pressure to innovate leads to greater capacity to innovate.

This result shouldn't come as a complete surprise. The logic is similar to that explaining why baseball players in the majors have higher performance than those in the minors. A significant component of that is selection (survival of the fittest). Of the roughly eight million Little League players in the United States, only 853 (0.01 percent) make the majors, so the ones who do make it are extremely talented. Similarly for companies, the failure rate in competitive industries is very high, so entrants have to be very good to survive.

But a sometimes overlooked component of competition is that it makes people stronger—the risk of being sent down to the minors provides constant pressure for players to maintain and improve their skills. Similarly for companies, the threat of losing share to rivals forces them to innovate.

PUTTING THE INSIGHTS INTO PRACTICE

Now that you know competition is healthy for RQ, what can you do about it? You're already in an industry, and its competitive structure is already in place. While it may seem companies have no control over the level of competition they face, that's not actually true. Companies shape the level of the competition in their industry on a regular basis. The recent intra-industry mergers attest to that: American Airlines and US Airways, Dow and DuPont, AB InBev and SABMiller, Kraft and Heinz. The problem with these mergers is they are moving in the wrong direction for innovation. They're moving to old

Porter rather than new Porter structures. We saw in Chapter 1 what happened to GE when it did this—it killed its innovation engine, just as Ron and Dan's theory predicts. So one strategy to support higher RQ is to resist industry consolidation.

Another strategy, one we almost never see once the founder leaves, is to innovate "ahead of the competition"—innovate to prevent potential rivals from entering your market—to enjoy the fruits of monopoly without paying the price of reduced RQ. Dolby Laboratories, the audio/video signal processing company, is a stellar example of this.

Dolby Labs is a $980 million company that develops, manufactures, and licenses audio/video signal processing technology. The company was founded in 1965 in Great Britain by Ray Dolby, an electrical engineer and physicist who contributed to the development of the first consumer VCR at Ampex, while working part-time during college.

At the time Dolby began his career, recording studios were faced with a choice of recording media. The traditional medium was recording directly to master disk. This preserved the quality of sound, but it was inflexible—any problems in the recording required the music to be replayed in its entirety. The new technology, analog tape, allowed recording of multiple tracks and facilitated editing by rewinding and recording over the flawed portion. The problem with the new technology was hiss.

Dolby developed "noise reduction" technology to reduce hiss by amplifying high frequencies (the desired sound) during recording such that they swamped the hiss, then reversing the process during playback. This created complementary products: those for the recording industries to amplify high-frequency sound, and those for the playback industries to de-amplify the entire track. In other words, the studios could generate recordings with noise reduction capacity, but in order to hear recordings without noise, listeners also needed compatible noise reduction technology in their playback systems.

Dolby's business strategy was to (1) manufacture products for the high value commercial sector, (2) generously license technology to the consumer product manufacturers at extremely low royalty rates (such that they had no incentive to design around the technology), (3) require that the manufacturers feature the Dolby logo on all licensed products, (4) aggressively court all elements of the value chain, including standards-setting bodies, (5) rather than resting on its patent monopolies, innovate at a rate that precluded competitors from capturing market share in a new generation of technology, and (6) only invest in technologies that could be protected by patent.

These policies are tightly interconnected. The policy of requiring patent protection (#6) both enables licensing (#2) and requires the company to innovate faster than the patents expire. The complementary products (recording plus playback) form an entry barrier in that new entrants need to develop two technologies and simultaneously capture two distinct markets. The policy of manufacturing for at least one market (the commercial market) ensures Dolby understands the manufacturing implications of its designs (an additional value-add to its licensees). Leaving the larger scale consumer market for licensees minimizes problems of obsolete inventory of the products and vintage capital of the associated manufacturing equipment.

Dolby has indeed been highly innovative under this strategy. It continues to generate patents at a rate of 10 applications per month. (In fact, its industry classification is "patent holder" rather than electronics manufacturer.) This innovation has allowed it to expand from its initial market of audio recording and playback for music studios into video recording and playback for consumers as well as in theaters (including 3D).

Indeed, the policies are highly effective. Dolby's policy of dual-sided markets and courting all elements of the value chain has ensured a near monopoly on noise reduction and related technologies. This monopoly was formalized when the Federal Communications Commission (FCC) adopted Dolby as the audio standard for digital television (DTV). Furthermore, its policy of requiring that recordings, products, and studios all display the Dolby logo made its brand ubiquitous without the need for advertising.

Together these policies ensure a high price premium. In addition, the policy of licensing to the larger market means capturing margins on products at large scale production, but without the need for capital investment. Accordingly the company's cash flow allowed it to grow while remaining private until Ray Dolby began worrying about estate taxes and chose to go public in 2004.

The only obvious misstep to threaten the company was Ray Dolby's assumption that studios would not convert to digital sound. This created the entry wedge for Digital Theatre Systems (DTS) to introduce digital sound technology for theaters in 1990. Within 10 years, more than 20,000 theatres worldwide had adopted its technology. While Dolby Labs lost its leadership in theater sound systems, its innovation engine allowed it to make a dramatic comeback with its superior digital technology, Dolby Digital (DD). That technology was introduced in theaters in 1994, and by 1998 the number of theaters in the United States with DD surpassed the number with DTS.

Some of the elements of Dolby's strategy are unique to its context—the only other company I can think of with "complementing technology" (where neither is valuable without the other) is Monsanto, with Roundup and "Roundup Ready" seeds. The element that could be widely adopted, however, is that of patenting all your innovations. This strategy requires you to release all the knowledge underlying the patent to the public. This opens the door for you to compete with a future clone of yourself. The realization you'll have to contend with your own knowledge in the future forces you to continue to innovate in the present.

An alternative but related approach to the Dolby strategy of competing with a future clone is one that forces you to compete with a current clone of yourself. While few companies would adopt this strategy if left to their own devices, IBM required Intel to license its technology to Advanced Micro Devices (AMD) in 1979 as part of IBM's contract with Intel for the personal computer (PC) microprocessor.

Intel and AMD were both "Fairchildren"—semiconductor companies founded by employees who left Fairchild Semiconductor. Intel was founded in 1968 by Robert Noyce, Gordon Moore, and Andy Grove; AMD was founded in 1969 by Jerry Sanders. While Intel began as a memory company, AMD initially manufactured peripherals, such as math coprocessors.

At the time IBM required Intel to license a second source, the company already had similar agreements with AMD on peripheral components. Accordingly, it was the obvious choice as the second source for the microprocessor. While the Intel/AMD agreement has resulted in protracted litigation that ultimately led to its being severed, the initial licensing allowed AMD to become a viable competitor to Intel. In the process, AMD has often produced new technologies that not only outperform Intel's technology, but do so at lower prices. Examples include the Am286 (1983) that had twice the speed of Intel's 80286, and the Athlon (2003) which was the first 64-bit x86 chip, and which Intel cloned to create the Pentium in 2004. While Intel maintains processor shares above 90 percent in the notebook and server markets, and above 80 percent in the desktop markets, AMD has moved more quickly into new areas such as mobile and games.

Thus the forced coexistence with AMD has kept Intel out of the monopoly trap we saw in Figure 3-2b. Moreover, Intel is particularly fortunate that AMD is its rival, as AMD's RQ is in the top 50 (and well above Intel's since 2004). Other rivals may not have been able to challenge Intel sufficiently to keep it innovative. Thus, a final strategy for avoiding the monopoly innovation trap is selectively licensing your technology to create a competitor.

Both the Dolby intended strategy of creating a future clone and the Intel imposed strategy of creating a current clone are examples of what Ursula Burns, chairman and CEO of Xerox, refers to as a good enemy: "Everyone needs what I call a good enemy. From a business perspective, the good enemy is the competition in the marketplace. The good enemy forces you to excel."[9]

SUMMARY

We've seen that Blue Ocean strategies (the modern-day version of Porter's Five Forces) lead to high profits, but low levels of innovation, and accordingly low RQ. The returns to these monopolists' innovation are low because ultimately they reach a point where they have brought all the profitable customers into the market. The remaining customers require either an extremely low price or an extremely high level of performance. Once the monopolist reaches that point, the cost to bring in additional customers exceeds the profits from those customers. The problem of course is that this lack of innovation makes the monopolist's market attractive to new entrants.

So for companies to remain innovative, the most straightforward strategy is to seek out Porter's "four diamond" markets. The reason companies in these rigorous environments continue to innovate is that their life depends on it. If they don't innovate, they will lose most of their customers to a rival that does. These rivals will offer products at either a lower price or higher performance. Moreover, while this "forced innovation" is costly to companies in the four diamond markets, it is not ruinous. Rather, it makes them stronger. As a result, they have higher RQ and accordingly higher monthly stock returns.

A second and related strategy is to resist the temptation to consolidate your industry—keep it a four diamond industry. In many instances, however, companies have little control over their market structure. Some industries are natural monopolies, for example. While such monopolists need to be vigilant that natural incentives don't work against their innovating, they need not be doomed. They can create competition for themselves to force innovation. This can take the form of creating future clones (as in the Dolby case), where the company both innovates *and* continues to enjoy monopoly profits. Alternatively, it can take the form of creating current clones (as in the Intel case).

MISCONCEPTION 3: Spending More on R&D Increases Innovation

It seems almost obvious that increasing R&D investment will increase innovation. In fact, this belief is so widely held that when the government wanted to increase innovation in 1981, it implemented R&D tax credits. The government's goal for innovation is similar to companies' goal. Both want growth. Companies want growth because it increases their market value; the government wants economic growth because it increases the number of jobs, the level of wages, and accordingly the standard of living for all its citizens. What's convenient about economic growth is that it's largely just the collective growth of all the companies in the economy. So the best way for the government to achieve its growth goal is to help companies achieve their growth goals.

THE THEORY LINKING R&D TO GROWTH

The intuition that technological innovation leads to growth has existed since at least Adam Smith's *Wealth of Nations*.[1] So it is interesting that the formal theory linking R&D to growth is fairly recent. That theory, called endogenous growth theory, was introduced in a 1990 paper by the economist Paul Romer.[2] The foundation of that theory is the production function, which links inputs (typically capital and labor) to output. The first extension of the production function linking it to innovation came from another economist, however. Robert Solow, in 1957, observed that U.S. output was growing faster than what you would expect from the growth in capital and labor. He concluded that there must be a missing input. He called this missing input "technological change."[3] Technological change essentially increases the amount of output from any given level of other inputs. A simple example of a technology increasing the output of employees is personal computers. PCs makes employees more efficient (except of course when they're using them for leisure, such as ordering gifts on Cyber Monday).

Solow's observation of substantial growth from technological change begs the question of where that technological change comes from. This is where Solow and Romer differ. Solow believes that it occurs "exogenously," meaning outside the productive activity of companies. In this view, knowledge essentially advances on its own, and companies get to take advantage of that advancing knowledge for free. In contrast, Romer believes that technological change comes from purposeful investment in R&D. In Romer's theory, R&D activity combines with existing knowledge to create new knowledge. This new knowledge then combines with capital and labor to make them more productive.

While Romer was working at the level of the economy, more recent theory has translated his theory into company-level versions that generate similar predictions for growth.[4]

The powerful benefit of having a formal theory (rather than merely intuition) is that it generates predictions you can test empirically. Testing these predictions allows researchers to determine if the theory is valid. What's exciting about the predictions from endogenous growth theory is that they include an expectation that the economy will grow even in the absence of population growth, so long as there continues to be investment in R&D. Moreover, the theory makes a very specific prediction that the rate of growth is determined by the number of scientists and engineers doing R&D. In particular, the theory holds that doubling R&D labor will double the growth rate of the economy, a prediction called "scale effects."

THE GROWTH PUZZLE

Here's the puzzle: R&D labor has increased 250 percent since 1971, yet GDP growth has at best remained stagnant over the same period. In fact, a better characterization is that GDP growth has actually declined 0.03 percent per year! The fact that one of Romer's predictions doesn't hold up suggests Romer's theory might be wrong.

Chad Jones, the economist who first documented the failure of the "scale effects" prediction in 1995, suggested the problem was that R&D has gotten harder, and Romer's theory doesn't account for that.[5] Jones proposed that two mechanisms contributed to this. The first mechanism is something he called the "fishing out effect." This is the notion that all the good ideas have already been cherry-picked. If you think about recent innovations such as personal computers, the Internet, and smartphones, you might be skeptical about this idea, but we'll examine it more concretely in a moment. The second mechanism is "diminishing returns to R&D labor," which is the notion that adding more scientists leads to substantial duplication of effort. Both of these ideas seem plausible. In fact, Robert Gordon, in *The Rise and Fall of American Growth*, makes similar arguments.[6] The dismal outcome if Chad Jones and Robert Gordon are correct is that growth will decline to zero (other than for population growth).

I have a more optimistic explanation, which is that companies have gotten worse at R&D. While companies getting worse may not *sound* more optimistic than R&D getting harder, if it's true, then we return to Romer's world of perpetual growth as long as there continues to be R&D. Since Romer's theory has been extended to the level of the company, this means not only will the economy continue to grow but companies should continue to grow as well. Moreover, if firms restore their prior level of R&D productivity, they should return to their prior levels of growth.

The big question now is whether Chad Jones and Robert Gordon are correct that R&D has gotten harder, or I'm correct that companies have gotten worse at it. We already saw in Chapter 1 that, on average, companies' RQs have declined 65 percent. This suggests I'm right that companies have gotten worse. However, it will also *look* like companies have gotten worse if in fact R&D has gotten harder.

So we need a different test of whether R&D has gotten harder. You need to think outside the box to come up with such a test, but one thought is that if R&D has truly gotten harder, it should have gotten harder for everyone. In other words, not only will average RQ decrease each year, but maximum RQ will decrease as well. Thus if I take the best company in each year (the one with the highest RQ) and compare it to the best company the following year, then the later companies will tend to have lower RQs than earlier companies.

That's not what I found when I examined 40 years of data. I found instead that maximum RQ was increasing over time! When you think about all the marvelous companies that have been created as part of the Internet economy, that seems plausible, but there is still reason to be skeptical when you're swimming against the tide. Next, I checked whether the same pattern held if, instead of looking at all public companies, I restricted attention to a particular sector, such as manufacturing or services. I found that maximum RQ was increasing within sectors as well. I then looked at coarse definitions of industry using the U.S. Department of Labor's Standard Industrial Classification (SIC) codes, such as Measuring Equipment (SIC 38), then successively more narrow definitions, such as Surgical, Medical, and Dental Instruments (SIC 384), then Dental Equipment (SIC 3843). What I found was that as I looked more narrowly, maximum RQ was decreasing over time (Figure 4-1). Thus Chad Jones's theory might hold at the industry level. Within a given industry, R&D does appear to be getting harder over time. However, what's exciting is that as some industries are dying off, companies are creating new industries with more technological opportunity. So overall, R&D appears to be getting easier rather than harder! It is easy to think of relevant examples: brick-and-mortar retail is dying, but Internet retail sales are growing; landlines are dying, but smartphone innovation seems to know no bounds.

To summarize where this leaves us with respect to endogenous growth theory, it appears that Romer's theory is left intact. While his scale effects prediction doesn't appear to hold (GDP growth is stagnant despite substantial increases in R&D labor), this seems to be due to a decline in companies'

FIGURE 4-1. **Average annual change in maximum RQ**

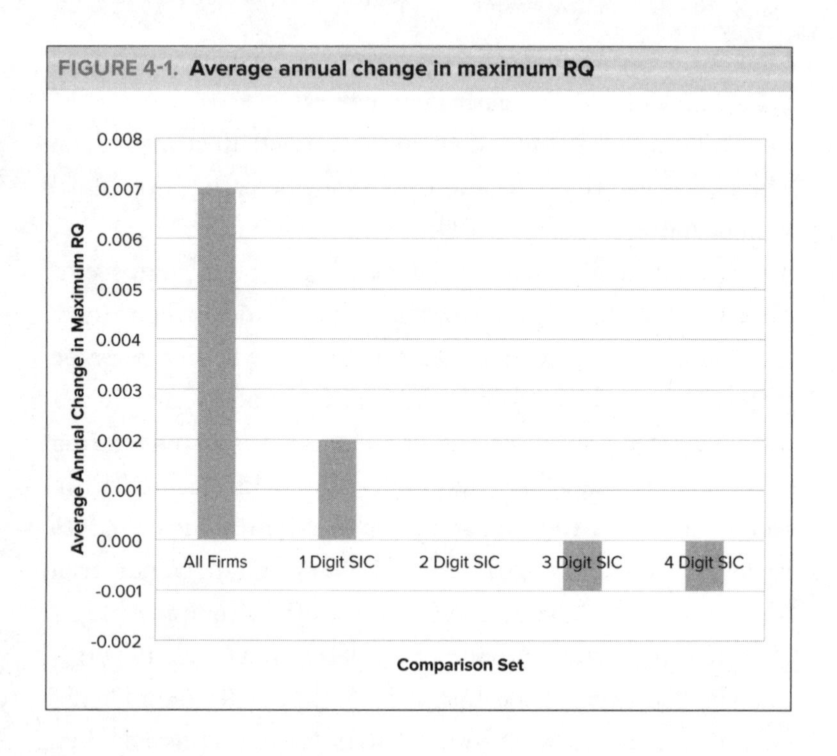

R&D productivity (companies getting worse) rather than from a decline in technological opportunity (R&D getting harder). The encouraging implication from this is that companies and the economy should be able to grow in perpetuity as long as companies continue to conduct R&D. The less encouraging implication is that this won't happen if RQ continues to decline. The challenge then is to determine *why* companies have gotten worse. If we understand that, then there's hope they can regain their prior RQ, and accordingly restore their higher prior growth. If enough companies do that, then the economy should also return to the pre-1970 growth levels that Gordon hails in his book.

Now that I've introduced some new theory, it's of course time for a test. What do you think happened when the U.S. government introduced R&D tax credits as part of the Economic

Recovery Tax Act of 1981? In answering the question, it may be helpful to remember that the goal of the act was to reverse the dramatic decline in companies' R&D investment that began in 1964. The hope was that restoring R&D investment would revive economic growth and bolster U.S. competitiveness against the then rising threat from Japanese manufacturing. Accordingly, this is a two-part question: (1) Did R&D tax credits restore R&D investment? and (2) Did the tax credits revive economic growth?

Now for the answers. The answer to question 1 is that the tax credits were extremely effective in increasing R&D. Within four years of implementation, R&D investment (plotted as diamonds in Figure 4-2) was restored to within 10 percent of its 1964 peak (2.9 percent of sales). However, the answer to question 2 is that GDP growth (plotted as squares in Figure 4-2), which used to follow R&D spending with about a three-year lag, never responded to the increase in R&D investment.

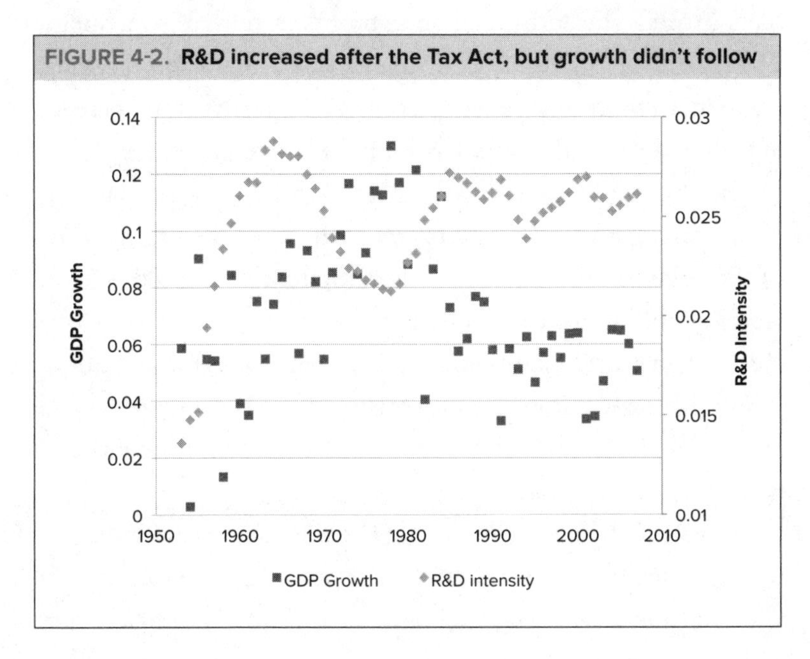

FIGURE 4-2. R&D increased after the Tax Act, but growth didn't follow

Why did tax credits win the battle of restoring R&D, but fail to win the war of restoring growth? For the same reason we just discovered when we investigated the failure of the scale effects prediction: companies were getting worse at R&D—their RQ was declining. Figure 1-1 from the first chapter made this point vividly—the rise and fall in RQ and GDP growth lined up closely with one another. Thus it's clear that GDP growth tracks R&D productivity rather than R&D spending.

AN ADDITIONAL PROBLEM

The failure of R&D tax credits to achieve their ultimate goal of growth is not the only problem with them, however. An additional problem is that tax credits caused a number of companies to *overspend* on R&D. What I mean by overspending is levels of R&D that decrease profits because the additional R&D investment exceeds the additional gross profits from that investment. Overspending on R&D continues to be a major problem for companies, for reasons beyond the tax credit. In fact, 63 percent of companies overinvest in R&D! How do we know this?

In Chapter 1 we learned that because RQ is based on economic principles, it is useful for more than benchmarking. One of the things it allows us to do is compute the optimal level of R&D spending for each company. What we mean by optimal R&D is the level of investment that generates the maximum profits. The calculation involves a standard piece of math called a partial derivative (the details are in Chapter 10). In essence, it's an exercise in marginal returns—determining the point at which an additional dollar spent on R&D begins to reduce profits.

Let's see how this analysis plays out using the example of Callaway Golf. Callaway Golf is a sporting goods company that grew from Ely Callaway's acquisition of Hickory Sticks in 1984. From its inception, Callaway was committed to advancing technology that made golf equipment more forgiving. Some notable product introductions were the S2H2 iron (1988) that relocated weight from the hosel to the clubhead for ease of launch, the Big Bertha driver (1991) that increased the distance and accuracy of off-center hits, and the Rule 35 golf ball (2000) that used aerodynamicists hired from Boeing to combine all the performance benefits of distance, control, spin, feel, and durability in one ball. In the past balls specialized in one of these dimensions, so players had to choose a ball that excelled in the dimension they most needed, while giving up performance along the other dimensions.

Callaway's commitment to technology is reflected in its R&D investment—its R&D intensity is 50 percent higher than the average for the industry. Further, its investment is smarter—its RQ is 16 points above the industry average. Most impressively, and most relevant to our current discussion, Callaway's R&D investment is very close to optimal. Callaway is among the 4.6 percent of companies whose R&D investment is within 10 percent of its optimum. In short, Dr. Alan Hocknell, senior VP of R&D for Callaway, seems to have great intuition for directing Callaway's technology efforts—he both generates greater revenues from each dollar of R&D and knows almost precisely when an additional dollar of R&D has insufficient payoff.

Despite that, even Callaway could improve its profits by slightly adjusting its R&D. While 2015 R&D of $31.3 million translated into 2015 net income of $14.6 million, Callaway could have earned even higher profits had it invested $32.7 million in R&D. Note of course this assumes Callaway had $1.4 million in projects of comparable quality that it could add to its portfolio. This might be a stretch for a small company with only a handful of projects, but is much more plausible in large multi-divisional companies.

HOW R&D BUDGETS ARE SET CURRENTLY

Using RQ to identify the optimal R&D is how companies should determine their budgets, because it maximizes their profits. In the past, however, companies wouldn't know their RQ, so it would have been almost impossible to do this. Instead they typically undertake a months-long process each year that matches top-down financial targets from the CFO's office to bottom-up requests from divisions and central labs to fund specific projects. To oversimplify, these bottom-up requests are rank ordered and approved up to the point where their combined cost crosses the financial target. A key question then is how the financial target is established. In general, the target applies the company's historical R&D intensity (R&D to sales ratio) to projected revenues for the upcoming year. Thus, if historically a company invests 5 percent of revenues in R&D and if projected revenues are $20 billion, then the R&D target is $1 billion. However, R&D intensity is a metric that is often benchmarked to industry rivals in an "arms race" to ensure that relative positions are maintained. So if a rival begins to increase R&D intensity, there would be pressure for the CFO to increase the company's own R&D intensity.

The problem with this approach is perceptively captured by George Hartmann, a former principal in the Strategy and Innovation Group at Xerox, who notes, "This approach implicitly assumes that the R&D budget follows revenue and profit growth, rather than driving it."[7] The approach treats R&D as an expense rather than an investment. While the bottom-up proposals include projected returns, these are used principally to rank order projects to determine which are funded versus rejected. This treatment of R&D as an expense is interesting because the dominant role of R&D in most organizations is to stimulate future growth.

So we've seen two different approaches to R&D budgeting: the rules of thumb approach of applying industry R&D intensity to forecasted sales, and the RQ approach of optimizing R&D. How prevalent are each of these approaches? While this may seem like a silly question given that companies haven't known their RQ, it's possible that company executives have very good intuition about when additional R&D will be profitable. In other words, even though they couldn't calculate the optimal level of R&D, they may have had a sixth sense that allowed them to home in on it. Callaway appears to be a good example of that.

To find out which approach to R&D budgeting companies were following, I created a simple test. To identify when companies establish R&D budgets using rules of thumb, I found the average R&D intensity (R&D divided by sales) in each industry for each year, then translated that into the level of R&D investment a company would have made. This merely multiplies the company's revenues for the year by the R&D intensity for its industry that year. To identify when companies were establishing R&D budgets by optimizing, I computed their optimal R&D in each year using the formula in Chapter 10. I then tested whether the rule of thumb or optimization was better able to predict the amount companies actually invested in R&D in each year. What I found was that the simple rule of thumb explained 79 percent of companies' R&D investment, while optimization explained less than 1 percent. (The remaining 20 percent of firms appeared to be using an approach other than rules of thumb or optimizing.) It seems clear that the majority of companies are relying on rules of thumb when setting their R&D budgets.

In principle using rules of thumb is a reasonable approach to tackling the problem of establishing R&D investment—just as it is for tackling a host of other problems. Rules of thumb typically reflect the collective wisdom from trial and error processes by lots of players. The problem with rules of thumb in the R&D context (versus an advertising context, for example) is that feedback from the R&D trial and error process doesn't occur until a number of years in the future. By that time, other things in the world have changed. Thus the feedback is not very closely linked to the trial.

A pertinent example of rules of thumb gone wrong because of long lags between R&D and revenues comes from the pharmaceutical industry. For most of the 1970s, pharmaceutical companies invested roughly 5.5 percent of revenues in R&D, which was near optimal for them (Figure 4-3). Thus they were another example of reliable intuition despite the unavailability of RQ.

After that however, the pharmaceutical world was rocked by four major shifts. The first shift was the Health Maintenance Organization Act of 1973. This led to consolidation of buying power and ultimately the creation of formularies that restricted the set of reimbursable drugs. This enhanced power to restrict reimbursable drugs translated into lower drug prices and/or smaller markets. The second shift was a technological shift in drug discovery from a chemistry foundation to a biological foundation that coincided with the 1982 introduction of Humulin, the first biotech drug to achieve FDA approval. This

FIGURE 4-3. **Pharmaceutical companies' rules of thumb matched optimal R&D until the late 1980s**

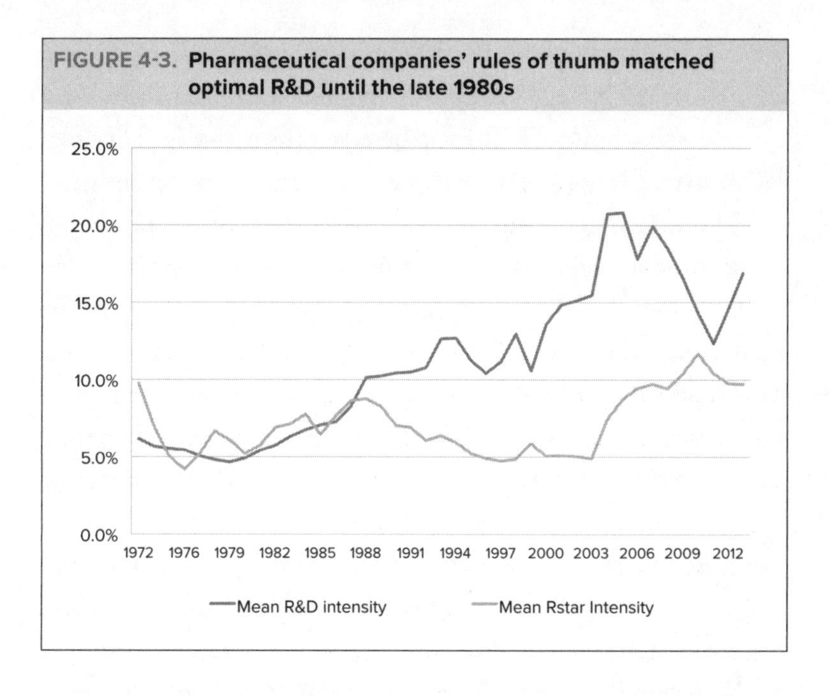

shift meant that the technical expertise companies had developed over decades would become obsolete. The third shift was the Hatch-Waxman Act of 1984 that paved the way for generic drugs to compete with proprietary drugs immediately after patent expiration. This meant that the commercial life span of drugs and therefore total profits from a given drug were substantially reduced. The fourth and final shift affected marketing. When the FTC lifted its moratorium on direct-to-consumer (DTC) advertising in 1985, marketing emphasis shifted from detailing to doctors to advertising to consumers. Thus, not only were companies' key technological resources becoming obsolete, so too were their marketing resources.

In response to these changes, pharmaceutical manufacturers began increasing R&D to about 8 percent of revenues. This trend toward increasing R&D intensity continued through 2004, reaching a peak at over 20 percent of revenues, four times what it had been in the 1970s! Unfortunately, pharmaceutical companies' RQs began declining in 1989, such that firms were overinvesting in R&D. The gap between the actual and optimal R&D over all those years became forgone shareholder returns.

The pharmaceutical industry is an extreme example. In other contexts the world changes more slowly. One slow change that seemed to affect most large public companies was the increasing amount of CEO compensation tied to stock price. This made R&D an easy target when companies faced quarterly earnings pressure. Since R&D is expensed rather than capitalized, cuts yield immediate increases in profit, while the detrimental impact of those cuts isn't felt for a few years. In marginal trade-offs between investments in physical capital or advertising whose returns are more easily quantified, R&D loses out. Evidence of this comes from companies' response to the recent recession. On average, companies with revenues greater than $100 million reduced their R&D intensity by 5.6 percent on average, whereas capital intensity fell only 4.8 percent and advertising intensity actually increased 3.4 percent.

How Accurate Are the Rules of Thumb?

The pharmaceutical example and the investor pressure example illustrates what can go wrong with rules of thumb, but four out of five companies still use them, so maybe they're reasonably effective. To examine that, I computed for each company in fiscal year 2014 what its optimal R&D budget would be, then compared that amount to what the company spent. This is shown in Figure 4-4. Each dot in the figure represents a company. Drawing a vertical line from a company to the x-axis identifies that company's *optimal* R&D investment (on a logarithmic scale). Drawing a horizontal line from a company to the

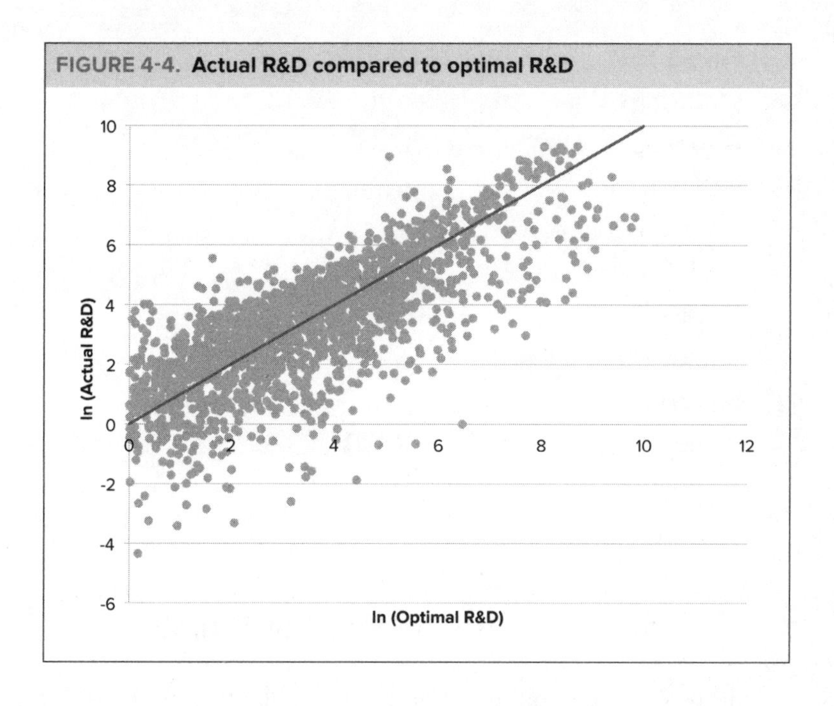

FIGURE 4-4. **Actual R&D compared to optimal R&D**

y-axis identifies that company's *actual* R&D investment (also on a logarithmic scale). Companies that are investing optimally fall on the diagonal line. There are almost no companies on that line! In fact, only 4.6 percent of companies are within 10 percent of that line in either direction. Furthermore, of companies that invest over $100 million in R&D, there were only 16 such companies, which appear on the following list. What's interesting about this list is that these companies don't seem to have much in common, with the possible exception that five of them (those in bold) are still run by their founders.

Abbott Laboratories

Alexion Pharmaceuticals

Amazon

CareFusion Corp

First Solar
FMC Corp
Goodyear Tire & Rubber
Halliburton
Johnson Controls
MSCI
Netflix
Rockwell Automation
Sandisk
Schlumberger
Teradata
Waters

How Costly Are the Rules of Thumb?

We now know companies should be establishing their R&D budgets by setting the marginal dollar of R&D equal to the marginal dollar of gross profits generated by the R&D. We also know that only about 4.6 percent of companies are very close to that level of investment. So rather than optimizing their R&D, companies appear to be using rules of thumb, and those rules of thumb are not very good proxies for optimizing. The next question is, how costly is it for companies to use industry rules of thumb when setting their R&D budget?

To examine that I looked at the impact on profits of increasing R&D 10 percent for the 33 percent of companies that were underinvesting in R&D, as well as the impact of cutting R&D to the optimum level for the 63 percent of companies that were overinvesting. For the companies that were underinvesting, a 10 percent increase in R&D would increase profits $36 million on average. (I capped increases at 10 percent because growing R&D beyond 10 percent in a single period typically strains the organization.) For the companies that were overinvesting, reducing R&D by the amount of overinvestment would increase profits $258 million on average (note this average is driven by outliers).

The nice thing about solving the overinvestment problem is that it shows up in profits right away, so it should have an immediate impact on the company's stock price. For the companies that are underinvesting, the gains from increased R&D won't be felt as profits for a few years. However, if the market correctly values the expected increase in profits, these companies should also see the gains in stock price immediately.

I first flagged the suboptimal R&D investment problem in a 2012 article in *Harvard Business Review* (*HBR*). There I looked only at the top 20 companies and found the combined market value those companies were leaving on the table was *one trillion dollars*! The claim that companies were investing this poorly was bold, and not surprisingly it received a lot of criticism by bloggers at *Atlantic* and *Forbes* as well as in comments on the *HBR* article itself. The main complaint was that *companies couldn't be that far off.*

Were the critics right? Since the article is now a few years old, I was able to rely on the test of time to answer that question. I looked at the performance of those 20 companies two years later (long enough for companies to implement the recommendations and begin to see results). Three of the companies no longer existed under the same name, so I restricted attention to the remaining 17 companies. I interpreted the companies' decisions in terms of the direction of their R&D investment levels, rather than exact amounts—did they increase or decrease R&D investment? Of the 17 companies, 9 followed my recommendation, while the remaining 8 companies deviated from my recommendation.

The important question however was not what companies did, but rather what happened as a consequence of that. The nine companies that followed my recommendation showed an average profit *increase* of 16.4 percent. In contrast the eight companies that deviated from my recommendation showed an average profit *decrease* of 14.1 percent. This is by no means a statistically significant test. It involved only a handful of companies, all of whom have many other factors affecting their profits. However it does suggest the underlying prescriptions have merit.

So we know that underinvestment leaves $36 million of profits on the table each year, which is substantial in its own right. However, there is an even deeper problem with underinvestment. Prolonged underinvestment leads to deterioration in R&D capability. Starving the company of R&D cuts meat as well as fat. We saw this happen with GE and HP, but they are the rule rather than the exception.

THE PEANUT BUTTER PROBLEM

We now know that 80 percent of companies use rules of thumb in setting their R&D budgets. We also know they appear to be leaving shareholder money on the table as a result of doing this. Finally, we know the correct approach is to use RQ to identify the optimal target from the CFO's office, and then to continue the bottoms-up approach to rank order projects and fund the highest-ranking projects until the target amount is reached.

The rank ordering of projects in this approach should ensure that the most valuable projects are funded. However, in multidivisional companies, one thing that might happen is that this rank ordering might produce a portfolio of R&D projects confined to a single division. This would occur if all projects in that division had higher expected returns than those in other divisions.

Typically the company would reject such a portfolio, since it wants all divisions to grow. Accordingly, in multidivisional companies another layer of targeting takes place. In addition to a single top-level target for the company, the CFO will apply industry specific benchmarks for R&D intensity to each division. This is what one CTO referred to as the "peanut butter approach" to R&D budgets. The problem with the peanut butter approach is that divisions differ in their RQ.

Across the companies I've worked with, the range of RQs across divisions has been as high as 30 RQ points (from the bottom sixth to the top sixth of companies). Since these data are highly confidential, let's create a fictional example of what might happen in such a company (Table 4-1).

TABLE 4-1. Fictional example of R&D budget in multi-divisional firm

	DIVISION 1	DIVISION 2	DIVISION 3
RQ	95	105	125
Revenues	$4000M	$6000M	$3000M
R&D	$160M	$240M	$120M

There are three divisions in our fictional company. Division 2, the largest division, has $6 billion in revenues and an above-average RQ of 105; Division 1 has $4 billion in revenues and a below- average RQ of 95; while Division 3, the smallest division, has revenues of $3 billion and a near-genius RQ of 125. Currently, the CFO allocates 4 percent of projected revenues to R&D for each division, so Division 1 gets $160 million, Division 2 gets $240 million, and Division 3 (which has the most productive R&D) gets the smallest amount—$120 million.

Let's compare two budgeting scenarios for the upcoming year, both of which assume top-level growth of 5 percent, so the R&D budget increases 5 percent ($26 million). In the first scenario, the CFO once again follows the peanut butter approach, so Division 1 gets an additional $8 million in its R&D budget, Division 2 gets an additional $12 million, and Division 3 gets an additional $6 million (the light columns in Figure 4-5a).

In the second scenario, the CFO allocates the additional R&D based on the marginal returns to R&D in each division, so Division 3 gets the entire $26 million in new R&D (the dark columns in Figure 4-5a).

The big question of course is what impact each of the two approaches has on revenues and profits. Looking first at the peanut butter approach, the additional $8 million in R&D to Division 1 yields $17 million in new revenues, the $12 million to Division 2 yields $44 million in new revenues, and the $6 million to Division 3 yields $40 million in new revenues. The total is an impressive $101 million in new revenues. If we assume gross profit margin is 50 percent, then the net profit from the new R&D is $24.5 million ($50.5 million in new gross margin; minus $26 million in new R&D).

Next let's examine the RQ approach of allocating the R&D based on the marginal returns. Because there is no new R&D to Division 1 or Division 2, they have no new revenues from R&D. However Division 3, which obtains all the new R&D because of its high RQ, yields new revenues of $172 million. This is 70 percent higher than the revenues from the peanut butter approach! Making the same assumption about margins, the RQ approach yields $60 million in new profits ($86 million in new gross margin; minus $26 million in new R&D). This is two and a half times greater profits than the peanut butter approach!

FIGURE 4-5. Comparing optimal allocation of R&D to the peanut butter approach

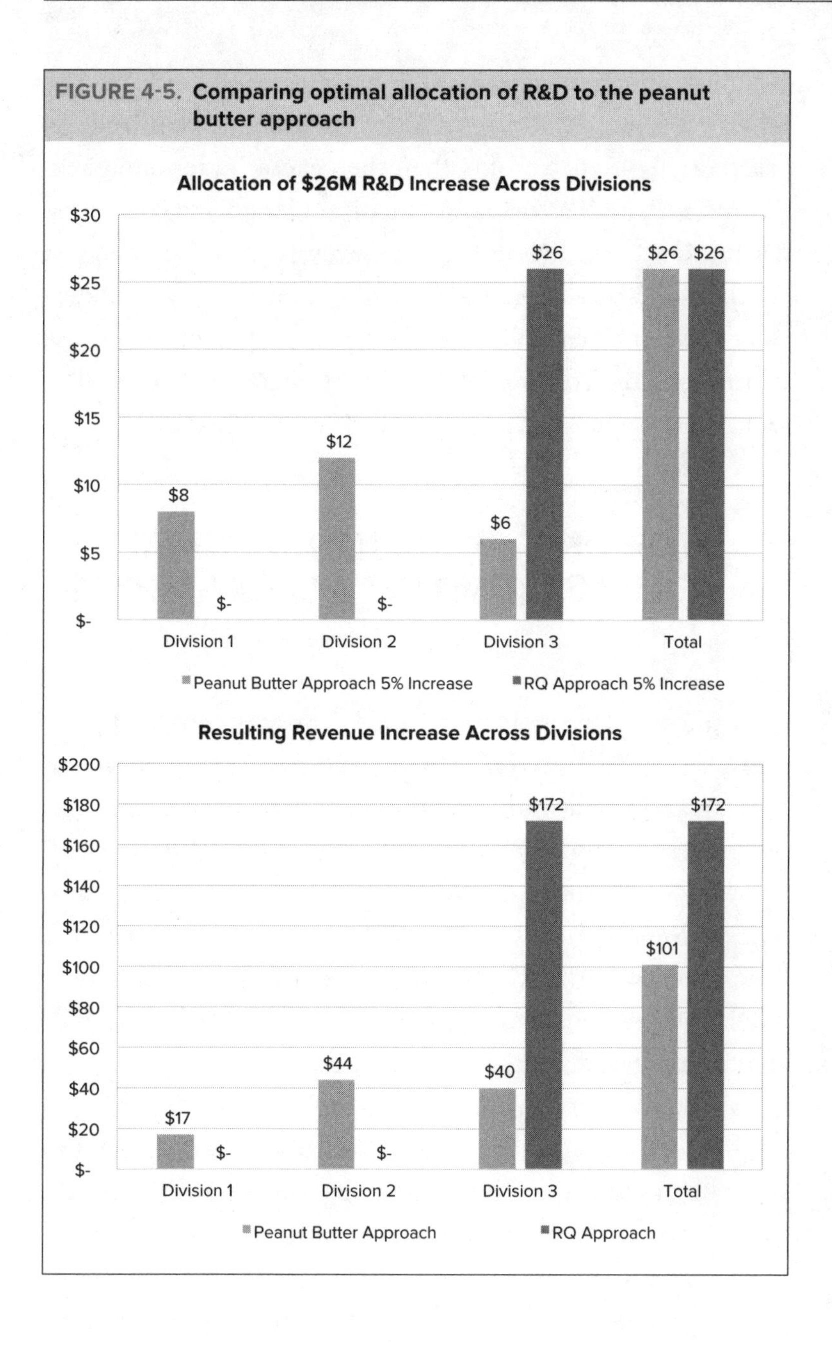

Thus, one tremendous and immediate benefit of using the RQ approach of allocating R&D to divisions is that it yields substantially higher profits than the peanut butter approach. However, there is also a more important long-term, yet subtle benefit. Once the divisions know they are competing with one another for R&D based on their RQ, they will try to understand the factors contributing to Division 3's higher RQ. Those efforts should lead to increases in RQ in the other divisions, which will lead to even higher profits.

WORKING THROUGH OPTIMAL INVESTMENT WITH TWO EXAMPLES

Let's see what optimal investment looks like in two types of companies—an example from the 63 percent of companies that overinvest, AMD, and another example from the 33 percent of companies that underinvest, McKesson. Both companies are interesting examples because they are in the RQ50, the 50 companies with the highest RQs among the set of public companies investing at least $100 million in R&D each year. (We briefly review all the RQ50 firms in the Appendix.) Thus these are R&D exemplars in one sense, and yet even they have trouble identifying the appropriate levels of R&D.

We learned about Advanced Micro Devices (AMD) in the last chapter when we discussed its role in pressuring Intel to innovate. In fiscal year 2015, AMD reported $1.1 billion R&D investment. That's substantially more than it should be investing. Given AMD's RQ, a 10 percent increase in R&D should increase revenues to $115 million. This sounds great, but the problem, given AMD's financials, is that the additional R&D ($107 million) *exceeds* the expected additional profits from that investment ($34 million). So AMD should shed projects that are ranked just below the $1.1 billion CFO target. If AMD does that, the company's profits should increase by the amount R&D is reduced.

AMD is a single example, but 63 percent of companies are in the overinvestment boat with it. While getting these companies to spend optimally actually reduces R&D, it generates $215 billion in higher profits that could be redeployed more effectively.

AMD offers insight for the companies that overinvest in R&D. Now let's look at the opposite problem, which is also prevalent—underinvestment in R&D. Here we'll use McKesson as an example. McKesson is an outstanding company in many regards, only one of which is being in the RQ50. It is one of the oldest companies in the United States (founded in 1833). Very few companies live that long, and when they do they tend to become lumbering. Instead, McKesson is ranked 11 in the Fortune 500 and is the largest pharmaceutical and medical supplies distributor in North America.

However, for all its virtues, McKesson, like AMD, is investing suboptimally in R&D. The company reported $392 million R&D investment in 2015. This is a decrease of $64 million from 2014, but it should have been moving in the opposite direction. Given its RQ, increasing R&D 10 percent ($39.2 million) should increase revenues $7.5 billion! Even with its relatively slim operating margins (6.1 percent), that still yields $417 million in incremental profits. At the current P/E of 15.7, shareholders would increase their wealth $6.5 billion if McKesson increased its R&D 10 percent. We don't recommend investing beyond that for two reasons: First, if R&D increases too dramatically, RQ will decrease. In addition, dramatically increasing resources typically leads to adjustment costs as the company tries to bring those resources online and integrate them with existing activities.

Again, McKesson is just one example, but 33 percent of companies are in the underinvestment boat with it. If all publicly traded underinvestors increased their R&D 10 percent, that would yield $16 billion in new profits.

RETURNING TO HP

In Chapter 1, I used the example of HP to demonstrate two problems that arise when companies lack reliable measures of R&D capability: (1) knowing whether HP's capability had deteriorated and, if so, by how much, and (2) knowing whether 2 percent or 9 percent of sales was closer to the correct level of R&D investment. I addressed the first problem in Chapter 1. I showed that HP's RQ had deteriorated substantially once neither founder was CEO. However, HP's RQ began to increase under Mark Hurd.

Now I address the second issue of the correct level of R&D investment. It turns out that because HP's RQ was declining, its optimal R&D intensity was also declining. Optimal R&D intensity increases with RQ because higher RQ companies get more bang from their R&D buck. Accordingly, it is optimal for them to spend more. In the case of HP, optimal R&D intensity fell from a high of between 5 and 6 percent under Dave Packard to a low of 2 percent when Mark Hurd became CEO. Thus Dave Packard was actually overinvesting in R&D, while Mark Hurd was investing optimally. So it is not the case that Mark Hurd was killing the innovation engine by cutting R&D. Rather, he was appropriately cutting the R&D investment because the engine had already deteriorated.

SUMMARY

Now we understand why tax credits won't solve the problem of increasing innovation and growth. They are the government's peanut butter approach to stimulating R&D. First, while the tax credits are very effective at increasing R&D, R&D in and of itself doesn't increase growth. This is in part because RQ has been declining, but also because 63 percent of companies are already overinvesting in R&D. Increasing their R&D even more will further decrease profits. On the flip side, the 33 percent of companies that are underinvesting shouldn't need tax credits to increase R&D. Increasing R&D will increase their profits even without tax credits.

We've shown that a better way to increase innovation and growth than the peanut butter approach of tax credits is using RQ to identify optimal levels of R&D, then increasing R&D in the companies and divisions that are underinvesting and decreasing R&D in the companies and divisions that are overinvesting.

What's exciting about this approach is that everyone is better off. Shareholders in companies that are overinvesting increase the value of their stock by having those companies cut overinvestment. Shareholders in companies that are underinvesting increase the value of their stock by having those companies increase their R&D 10 percent. Further, because both corrections increase profits, the government gains corporate income tax on the higher profits and capital gains tax on the additional shareholder wealth. So the government increases revenues *and* saves the tax credits.

5

MISCONCEPTION 4: Companies Need More Radical Innovation

Radical, or "new to the world," innovations are tremendously important to the economy. Successful ones completely transform our lives in such a way that it's hard to imagine life without them. Some notable examples include electricity (which replaced steam power and gas lighting), personal computers (which were so revolutionary, the closest thing we can think of their replacing is typewriters and calculators), and smartphones (which not only replaced landlines but also eliminated the need for lots of devices we used to own separately: answering machines, alarm clocks, GPS devices, cameras, video cameras, voice recorders, MP3 players).

Because radical innovations are so important, investors often chastise companies for not introducing more of them, as captured in a May 2013 article in *Bloomberg*: "Once known for inventing new categories, P&G hasn't had a $1 billion hit since re-imagining the household mop with the Swiffer a decade ago."[1]

THE RECORD ON RADICAL INNOVATION

The fact of the matter is that the prescription to generate more radical innovation is ill-advised. While radical innovations are tremendously beneficial to the economy, they tend not to be beneficial to companies. Radical innovations have lower returns than incremental innovations: RQ *decreases* in the percentage of sales from "new to the market" products. Conversely, RQ *increases* in the percentage of sales from incremental innovation. Moreover, the larger the company, the greater the returns to incremental innovation, and thus, the greater the penalty for radical innovation. This explains the tendency for radical innovation to decrease as company size increases—something we've known since Edwin Mansfield's work in 1981 examining the relationship between firm size and innovative activity.[2]

One of the reasons radical innovations have lower returns is that the market for them doesn't exist yet. This means the inventing company not only has to develop the new product, it needs to develop the new market as well. In many cases this requires convincing people they need something they've never needed before. This is extremely difficult. The markets for some inventions take decades to materialize. Television took 22 years from invention to commercialization, and many more years to become widely adopted. Even 20 years after their introduction, only 50 percent of U.S. households had personal computers.

A second reason radical innovations have lower returns is that the company that creates the market for them often loses out to later entrants who capitalize on weaknesses in the pioneer's initial design. Follower companies have the advantage of waiting to see how the market reacts to the initial product and then providing a version that better matches customers' preferences. Thus the followers capture a larger share of the market. Moreover, the followers do so with much lower investment because they get to free-ride on the R&D and marketing of the pioneer.

While there are no hard statistics, people who've tried to identify market winners tend to find they are more likely to be followers than pioneers. While Polaroid pioneered and dominated the instant camera market, and G.D. Searle pioneered and dominated the sweetener market, there are many more examples of winning followers. EMI invented and created the market for CT scanners but lost out to GE; Netscape pioneered the Internet browser but lost out to Microsoft's Internet Explorer; Sony pioneered the digital reader but lost out to Amazon's Kindle; and Diamond pioneered the MP3 player with its Rio but lost out to Apple's iPod.

A third reason radical innovations have lower returns is that they are less likely to capitalize on companies' existing resources. Because radical innovations typically serve markets outside the ones the company currently serves, they often require entirely new sales and marketing teams. In addition, they may require different manufacturing processes as well as servicing systems. This was the problem Raytheon confronted when it invented the microwave oven.

Raytheon had historically manufactured electronic components since its founding in 1922, beginning with rectifiers, then vacuum tubes, magnetrons, and ultimately whole radar systems during World War II. In 1945, Percy Spencer, while working on one of the company's radar systems, noticed that its microwaves melted a candy bar in his pocket. By year's end, the company filed a patent for a microwave cooking process, and in 1947 the company introduced a commercial version of the Radarange (five feet, five inches tall and 750 pounds!). The problem Raytheon confronted with the new microwave was diffusion. Prior to the launch of the Radarange, Raytheon had only manufactured industrial products (its primary customer was the U.S. government). It had no capacity to market, distribute, and service consumer products, so wisely, it chose to license its technology to Tappan Stove in 1952. Tappan introduced a home unit in 1955, but the unit was unsuccessful in the marketplace. It was another 10 years before Raytheon took another stab at the market by acquiring Amana in 1965. This acquisition resulted in the first successful consumer microwave oven—the countertop Radarange. So, while Raytheon ultimately became a rare pioneer winner, it took 20 years to reach the market with a viable product and required a failed experiment with licensing followed by acquisition of a consumer appliance manufacturer.

THE RADICAL INNOVATION DILEMMA

Here's the problem we face: the economy benefits tremendously from radical innovation, but companies' returns to radical innovation are negative. Given that, how are we going to get an adequate supply of radical innovation to benefit the economy?

The exciting answer is that the source of most radical innovations is basic research, and fortunately the returns to basic research are positive. Moreover, the returns to basic research increase with company scale. Accordingly, while large companies are less willing to do radical innovation, they *are* willing to do basic research—the source of radical innovation.

To understand why that's true, we first want to understand what basic research is. While you were introduced to it in Chapter 2, it plays a larger role here, so I want to remind you of its definition. The NSF defines *basic research* as the planned, systematic pursuit of new knowledge without specific immediate commercial application. It is distinguished from *applied research* (planned, systematic pursuit of new knowledge aimed at solving a specific problem or meeting a specific commercial objective) and *development* (the systematic use of research and practical experience to produce new or significantly improved goods, services, or processes).

Universities and government (and other nonprofit) labs conduct the bulk of basic research in the United States (53 percent and 27 percent, respectively), precisely because they don't need to commercialize or make a profit from their research. This concentration of basic research in nonprofits is a good thing, because it means the research is publicly available for later researchers to build upon.

Having said that, 20 percent of basic research is still done by companies. Moreover, this research is concentrated in large companies, because the returns to basic research increase with company scale. Richard Nelson argues that large companies have two advantages in doing basic research.[3] First is a capability advantage—they typically have a broader technological base, which means they are better able to make connections across domains that lead to new breakthroughs. Second is an incentive advantage. Large companies typically participate in a broader range of markets. This increases the odds that discoveries emerging from basic research will have an application somewhere within the company.

PROJECT PORTFOLIOS AND SUCCESS CURVES

So, large companies conduct most basic research that identifies opportunities for radical innovation. However, we saw previously that they are less likely to exploit that research as radical innovation, because the returns to radical innovation are lower than those for incremental innovation. So what happens to these ideas they generate but don't exploit?

To understand what happens to these abandoned ideas, we need to introduce two concepts: product pipelines and success curves. These two concepts are basically opposite sides of the same coin. Rather than define them, it's probably easiest to explain both concepts by way of an example from the pharmaceutical industry.

Pharmaceutical R&D begins with "discovery"—the screening of compounds for therapeutic potential. The rule of thumb in the industry is that it takes 5,000 of these compounds to generate a single FDA approved drug. The winnowing down of the compounds is captured in the product pipeline (columns) in Figure 5-1. Compounds that appear to have promise are advanced to preclinical development, a roughly 18-month process to evaluate dosage and safety. Of the 5,000 compounds that are screened, only 5 percent (250) are deemed promising enough to warrant this preclinical investigation. Even at that narrowing, the average out-of-pocket cost to take a compound through preclinical investigation is approximately $1.7 million per compound.[4]

The most promising compounds that survive preclinical development enter clinical testing. Clinical testing is a three-phase process lasting approximately five years, which proceeds from safety trials in a limited sample of subjects to efficacy testing in a larger sample. Of the 250 compounds that enter preclinical development, on average, less than 7 advance to clinical trials, and of those, only 1.5 survive all three phases of the trials. One thing limiting the rate at which drugs enter FDA trials is the fact that the average out-of-pocket cost is $965

FIGURE 5-1. Pharmaceutical product portfolio and success curve

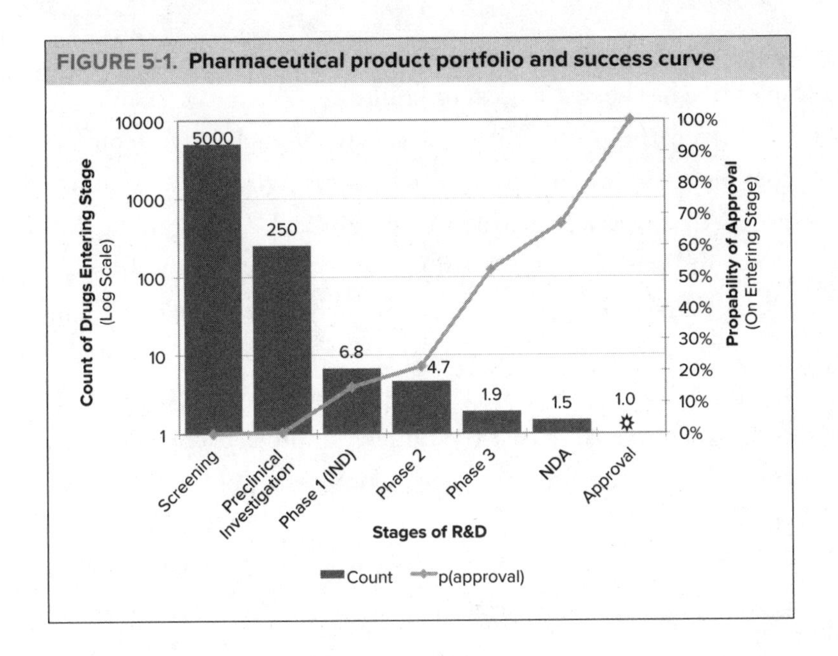

million.[5] Drugs that survive all three phases of clinical trials are submitted for FDA approval, a process that can take an additional two months to seven years.

The probability of surviving each stage is captured in the "success curve" (line plot) in Figure 5-1. At each stage of the process, the company learns more about the technical feasibility as well as the market prospects of a drug. This new knowledge on both the technical and marketing fronts can be translated into expected revenues, which can in turn be compared to expected future costs. If companies had unlimited budgets, any drug for which the present value of revenues exceeds the present value of costs would be funded through the next stage. However, companies don't have unlimited budgets, so they engage in a sophisticated process of determining which drugs to advance to each stage. Note the term "success curve" is a misnomer, because in most cases the company is *deciding* at each stage which drugs to fund in the next stage, rather than learning there has been a success or failure.

Given the probabilities of surviving each stage, companies maintain product portfolios comprising 5,000 compounds under discovery, 250 drugs in preclinical development, 6.8 drugs in Phase I trials, 4.7 in Phase II trials, and 1.9 drugs in Phase III trials for each drug it ultimately hopes to release. Thus the product portfolio (the columns in Figure 5-1) is essentially the inverse of the success curve (the line plot in Figure 5-1). It is designed to ensure that the number of projects entering each stage compensates for the expected probability of surviving that stage.

The success curve for the pharmaceutical industry is particularly harsh, but we see a similar pattern if we look across all industries (Figure 5-2). On average, it takes 3,000 raw ideas to achieve a commercial success.[6] Knowing this, for each expected

FIGURE 5-2. **Average R&D project success curve across all industries**

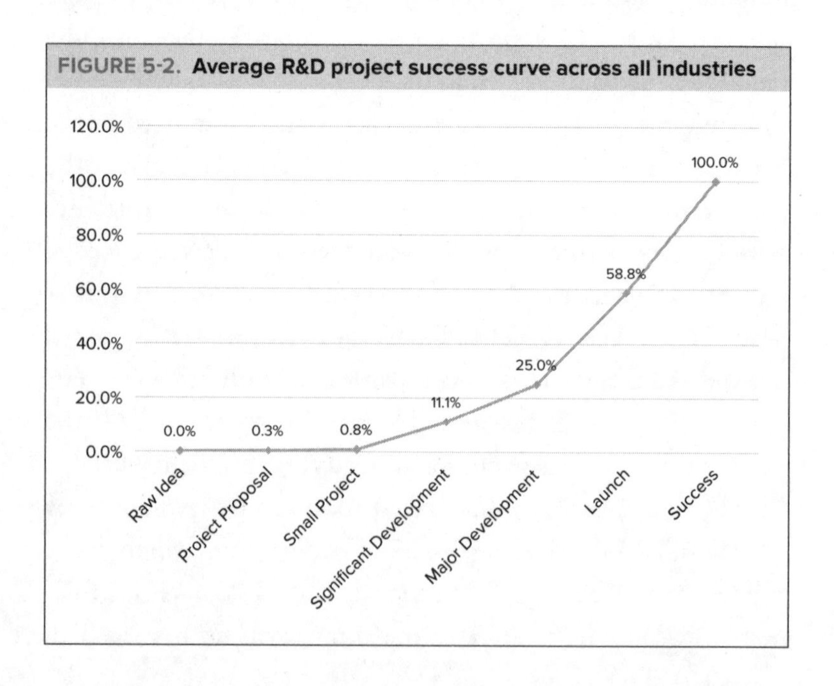

launch, companies try to generate 3,000 raw ideas, fund 125 of those at a preliminary level, fund 9 significant developments and 6 major developments, and launch 1.7 products.

Interestingly, this success curve for R&D project portfolios is almost identical to the success curve for venture capital portfolios—if we consider business plans they receive, review, and fund at various levels. Thus, the severity of the selection process for internal markets within large companies is comparable to the selection process of external markets for new ventures. Accordingly, large companies are not over-abandoning projects relative to the rate at which VCs abandon new ventures. If anything, both VCs and large companies pursue innovations too long. An interesting study by Isin Guler showed that if VCs could pull the plug more optimally, their returns would triple![7] So the answer to the question of what happens to all the ideas from basic research is that most of them are deliberately abandoned either because the forecasted revenues don't cover forecasted costs or because the company has a limited R&D budget and some projects lose out to others with higher expected returns. Thus there is a scrap heap of roughly 123 abandoned innovations per two launched innovations in each company.

WHY DO SMALL COMPANIES DO RADICAL INNOVATION?

Given all the disadvantages to radical innovation, why do small companies do it? The theory of why startups do radical innovation (while large companies do incremental innovation) comes from the economist Richard J. Rosen.[8] Rosen argues that because the output of large companies is greater, they generate substantial returns even if the improvements to existing products only allow them to increase prices for their products by a small amount. Accordingly, large companies prefer safer innovation that enhances their existing products. In contrast, precisely because they lack scale, small companies require riskier projects for which customers are willing to pay a substantial price premium.

So while Rosen answers the question of *why* small companies might be more willing to undertake radical innovation, he doesn't answer the question of *how* they do it. The answer is that they feast on the scrap heap of innovations abandoned by large companies.

One means to feast on the scrap heap is to "shop" for technologies, as in the case of Apple's "shopping" at Xerox PARC, which we discussed in chapter 2. We all know the story of Steve Jobs exploiting technology from Xerox PARC for the Lisa after touring its facilities. Xerox had recently made an investment in Apple and wanted to show Jobs what was going on in its lab.[9] As a result of the visit, Jobs committed to incorporate technology he saw there in the Lisa. Apple accomplished this in part by hiring PARC employees.

The more common path for feasting on the scrap heap is that employees from inside the company who contributed to development of the radical innovation leave to start their own company. This typically occurs because the team behind the abandoned invention has become committed to it and wants to see it come to fruition. This story played out at Hughes Aircraft Company in the context of laser technology.

While the theoretical underpinnings of the laser had been laid out by Albert Einstein in a 1917 paper, the race to implement the theory in a prototype didn't occur until a 1958 paper on optical masers.[10] The race included Bell Labs, RCA Labs,

Lincoln Labs, IBM, Westinghouse, and Siemens, but it was Ted Maiman at Hughes who in 1960 succeeded in demonstrating a working laser. Moreover, Maiman's laser was easy to build and thus had greater commercial potential.

The problem he confronted was a common one for radical innovations: "A laser is a solution seeking a problem," noted Maiman's colleague, Irnee d'Haenens.[11] So it isn't surprising Hughes refused to fund further development. Indeed, it had been reluctant to fund the initial $50,000 project. Accordingly, Maiman left Hughes and founded his own company, Korad, in 1961 to develop and manufacture lasers.

Thus what appears to be large company failure and small company success in developing radical innovation is in fact a broader system in which large companies generate inventions and make rational decisions to abandon them, and startups (most of whom are unsuccessful) attempt to commercialize some of the abandoned inventions.

What types of innovations do companies abandon? Since they abandon 123 out of 125, it's probably easiest to examine the ones they pursue, but let's take an extreme example of ones they are likely to abandon—flying cars.

Paul Moller has been developing a flying car for over 40 years. He even has a working prototype—a sexy cherry red VX400 that I saw at an auto show about 10 years ago. The VX400 has the footprint of a car, so it can park in a garage, and has vertical takeoff (like a helicopter or Harrier jump jet), so it can take off from a driveway. However, even after a $100 million investment, the car has only demonstrated flight while attached to a tether. One of the problems Moller faces is that the cost of developing a manufacturable model is enormous. The Cirrus Vision SF50 (probably the most comparable private jet) took 10 years from initial customer deposits in 2006 to the certification flight for its first production model in 2016, at an estimated development cost of $150,000,000.

However, development costs for the Skycar pale in comparison to the ecosystem challenges, a concept introduced in Ron Adner's book *The Wide Lens*.[12] In the Skycar case the most notable ecosystem challenge is getting the FAA to approve a new category of vehicles and provide flight rules to govern flights initiated from driveways. However, even if the enormous FAA ecosystem problem is solved, the next ecosystem problem is that "drivers" would require a pilot's license, something that costs thousands of dollars in flight time alone. Until these ecosystem problems are solved neither automobile manufacturers nor aircraft manufacturers will jump into the flying car fray.

Why are entrepreneurs willing to take on these projects that established companies rationally abandon? Passion and overconfidence. You can see the passion and determination of Ted Maiman for lasers in his autobiography, *The Laser Odyssey*,[13] and of Paul Moller for his Skycar on his company's website.[14] Moreover, both inventors truly believe they will succeed. Even when they know the odds of new company success are extremely low, they believe those odds pertain to "ordinary people," rather than themselves. Like the children in Garrison Keillor's fictional town of Lake Wobegon, they all believe they are above average.[15]

Thus while we as a society benefit tremendously from the contributions of these bold pioneers, these contributions come at substantial personal cost to the pioneers themselves. (We saw 98.75 percent of ventures fail to gain VC and then return their invested capital.) We rarely hear about those failures. So when we think about the returns to radical innovation we tend to compare the profits of the successes only to their own investment. What we should be doing is adding the investment of all the failed entrepreneurs to the denominator. This likely would present a returns number much lower than that of large companies who bear the 124 "failures" in their returns.

EXCEPTIONS TO THE "RULE"

Are there exceptions to the large company "rule" of avoiding radical innovation? Certainly. One example is Motorola's cell phone. Like Raytheon, Motorola began in the 1920s as a component manufacturer, but it moved quickly into radio manufacturing. It began TV manufacturing following World War II, becoming one of the top manufacturers. Motorola's success with televisions created such a substantial consumer market presence that when it launched a stereo phonograph in 1958, its sales that year exceeded those for the entire market the prior year.

So when Motorola demonstrated the first prototype of a cell phone, the DynaTAC, in 1973, it was well positioned to exploit that radical invention. In fact, the DynaTAC and its successor, the StarTAC, commanded as much as 60 percent of the U.S. market during their prime.

Why was Motorola an exception to the "large company avoiding radical innovation" rule? Because cell phones exploited valuable resources that Motorola had developed for its home entertainment products—large scale electronics manufacturing capability, a supply chain, a highly valued national brand in the closely related category of home electronics, a sales and distribution system to consumer electronics retailers, as well as a service network. In short, Motorola could free-ride on a lot of existing resources for mobile phones, whereas typically radical innovations require an entirely new set of resources (as we saw for Raytheon's Radarange).

HOW CAN WE REDUCE THE "WASTE" FROM ABANDONED PROJECTS?

Is there something we could do to reduce the "waste" from the scrap heap of 123 abandoned projects? When I discuss success curves with audiences, they express frustration that so many ideas are going to waste. One thing I explain to them is that this is a necessary cost to ensure that investment goes to the best ideas. That's not a particularly satisfying response to most people, so it's worth considering if there's something companies could be doing differently. After mulling this over for a number of years, I've been able to identify six strategies to better manage the scrap heap and thereby improve company RQ.

1. *Increase R&D funding.* This is the simplest strategy for exploiting the scrap heap, though it is only valid for the 33 percent of companies that are underinvesting in R&D. The easiest way for these companies to make use of the additional R&D is to fund the positive net present value (NPV) projects that fell just below the cutoffs at each stage of the pipeline. The logic here is that these projects are commercially viable, but the company isn't funding them because its underinvestment in R&D imposes funding thresholds that are too restrictive.
2. *Better manage the "chasm."* A second strategy is to develop better strategies for crossing the "chasm" between R&D stages—most notably the transition between central R&D (typically basic and applied research) and divisional R&D, though it also applies cross-divisionally. One example of research that might have ended up on the scrap heap without such mechanisms is the ion beam technology at Hughes Aircraft Company that we saw in Chapter 2. It

had been developed and abandoned by one group but was adopted by another. While this occurred serendipitously at Hughes, companies can and should create mechanisms for increasing the likelihood of such serendipity.

3. *Better manage the success curve.* This strategy seems to run counter to the first strategy, but is a natural outgrowth of the observation that people struggle with decisions to "pull the plug." As we mentioned previously, Isin Guler found that if VCs could do this more effectively, their returns would triple. Since R&D project portfolios' success curves match those for VCs, it's likely companies are pulling the plug too late as well. In research I did with Dan Elfenbein and Rachel Croson,[16] we found that one way to improve termination decisions was to take those decisions out of the hands of people who had a stake in their outcome—to have independent advisors make the decisions. Early termination of projects ultimately slated for the scrap heap changes the shape of the success curve so that projects are abandoned earlier when they have had only limited investment. This effectively reduces the amortized cost per successful project, thereby freeing up R&D funds for more projects at the front of the pipeline. So instead of funding 125 projects and obtaining 1 success, you might fund 180 projects and obtain 1.5 successes.
4. *Implement a recycling program.* This strategy seems to fly in the face of the powerful logic for abandoning projects, but there is a tremendously successful example from Xerox called Xerox Technology Ventures (XTV).[17] We discussed XTV in Chapter 2 as a way for large companies to exploit small companies' innovation advantages. Remember that XTV was an internal venture program established in 1989 to make better use of technologies from Xerox PARC.

While XTV had many of the features of traditional corporate venture capital (CVC), such as independent decision authority and carried interest compensation for the partners, it was unlike CVC in that it invested in projects that had been abandoned by PARC. Thus XTV was essentially a recycling program for the PARC scrap heap. Not only did XTV provide new life for abandoned projects, the fund generated a rate of return that was four times that of a typical VC fund (56 percent versus 13.7 percent). The XTV case demonstrates that tremendous value can be created from a recycling program. Moreover, XTV isn't the sole example of successful recycling of the scrap heap.

One of my favorite RQ50 companies, Medicines Company, is based entirely on a strategy of giving new life to other companies' failures. The company examines failed FDA trials of drugs from other pharmaceutical firms to determine whether the trial failed on the drug's merits or because of poor trial design or execution. Surprisingly, many failures are due to poor trials. In these instances, Medicines Company acquires the patent from the pharmaceutical firm at a bargain price (because it appears to have no residual value), and then redesigns and executes the new trial.

5. *Reduce the cost of development.* Early termination (strategy 3) is implicitly a strategy of reducing development cost as a way to fund more projects. It is possible to reduce development costs more directly. This seems like low-hanging fruit, but it is easier said than done. One way to do this is to reduce red tape. We saw in the case of the advanced workstation project at XTV that one advantage the project had at XTV versus mainstream Xerox was exemption from onerous requirements such as producing user

documentation in 38 languages. As a result of this and other XTV features, XTV was able to complete the project at one-sixth of Xerox's estimated costs. This suggests there is tremendous opportunity to cut project development costs within large companies.

6. *Establish knowledge markets.* The final avenue for better exploiting the scrap heap mimics what universities did in response to the 1980 Bayh-Dole Act. They established technology transfer offices (TTOs) to generate revenue from university technology. As we learned in Chapter 2, the principal mechanism for generating revenue was licensing to companies.

Of course, the key challenge in securing licenses is generating company interest in each technology. Some of this interest occurs because companies maintain links with university researchers in the domains related to their own research. They do this not only to stay on top of emerging research, but also to identify potential employees. However, the number of technologies generated at universities exceeds the number of industry relationships, so most universities established "knowledge markets" in which they post searchable databases of their patents, such as the following database for Washington University: https://otm.wustl.edu/for-industry/list-of-otm-technologies/.

While it is natural to think of these searchable databases as eBays for technology, eBay has at least two advantages over patent databases. First, buyers have a good understanding of what they are searching for. Second, sellers can use common language (as well as pictures) to communicate what they are selling. By way of comparison, here is a listing on Washington University's database: "Sulfide Trapped Antigenic Peptide in an MHC

Molecule." There is no description of the technology and no picture or diagram, though I'm not sure it lends itself to either. This is not my domain, but it's hard to imagine what keywords would generate this patent as a search result.

Some companies have already tried to establish technology markets for their abandoned technologies. A notable example is IBM during the period when James McGroddy was director of research.[18] Further, there are now established patent markets, which brought over $7 billion worth of patents to market in the past five years. However, the majority of these patents are sold in bundles, and in many cases are purchased as means to defend companies' existing technology from litigation, rather than to open doors to new technology.

Having said that, solving the problem of matching available technologies to potential users would be valuable even if the market existed only within the company. I remember one point while at Hughes discovering that another group within the company was developing technology almost identical to what my team was working on. Knowing that earlier would have led to either cost savings or greater advance on both sides. If companies can effectively solve the internal technology market problem, there is hope they could establish viable external markets for abandoned technology as well.

SUMMARY

Large companies, particularly those with central labs doing basic research, are going to generate the seeds for most radical innovations. These radical innovations will typically have lower returns than more incremental innovation, and thus will be part of the 123 of 125 projects the company abandons. In some cases, startups will exploit these innovations. Thus small companies will be disproportionately associated with radical innovation, even though the underlying technology for the innovation is likely to have originated in a large company. While most of those startups will fail to recoup their initial investment, those that succeed will have outsized returns.

This is a win-win-win system. Large companies get higher returns, entrepreneurs have a stock of abandoned innovations to exploit, and the economy gets radical innovations. But we might ask, "Is there a better solution"—is there a way for large companies to better capitalize on the 123 projects they abandon? I've provided six strategies suggesting the answer may be yes.

6

MISCONCEPTION 5: Open Innovation Turbocharges R&D

One of the biggest innovation trends over the past few decades has been "open innovation." The concept was first introduced in 1983 and later popularized by Henry Chesbrough in *Open Innovation: The New Imperative for Creating and Profiting from Technology*,[1] as well as related articles in the business press. The basic idea behind open innovation is that the amount of R&D done outside a given company will always dwarf what the company itself can do internally. Intel, for example, the largest U.S. spender, with R&D investment in 2013 of $10.6 billion, only represents 2.3 percent of the $456 billion total U.S. R&D. Accordingly, it is likely that research being done elsewhere is relevant to Intel. Given that, Intel may be able to extend the value of its own R&D by exploiting external research.

In order to define open innovation, it's useful to understand its antithesis, closed innovation. Chesbrough defines *closed innovation* as the traditional system in which companies generate, develop, and commercialize their own ideas. To accomplish this, they hire the best and the brightest, and as a result can generate a substantial number of high-quality innovations. Profits from the new innovations are reinvested in R&D, thereby creating a "virtuous cycle of innovation."

In contrast, Chesbrough defines *open innovation* as a system in which companies combine internal and external means for generating, developing, and exploiting innovations. This allows companies to use the best path (internal or external) for development and commercialization of their own ideas, as well as use the best ideas (internal or external) to fill their development and commercialization pipelines.

Chesbrough argues that the closed innovation model began to erode toward the end of the twentieth century, in part because increased labor mobility made it more difficult for companies to control their intellectual property. The other factor contributing to erosion of closed innovation was dramatic growth in venture capital (VC). Venture capital facilitated the creation of startup companies as an additional avenue for these mobile employees to exploit the innovations they helped create within the large companies.

In Chesbrough's view this erosion meant internal R&D was no longer a source of competitive advantage. As evidence of this, Chesbrough points to the fact that startup companies can't afford to do basic research, yet despite that, they are able to unseat large established companies. As a particular example, he points to the fact that Lucent and Cisco competed directly in the same industry, yet had vastly different innovation strategies.

While it is true that the two companies ultimately competed in the same industry, this occurred only after the convergence of voice and data. Lucent came from the voice side. It was the equipment manufacturing business spun out of AT&T

Corporation in 1996, with a goal of achieving scale economies by selling equipment to AT&T's rivals. As a means to attract former rivals to purchase from Lucent, AT&T included Bell Labs and all its patents as part of the spinout. Not surprisingly, given its legacy, Lucent conformed to the closed innovation model. It sought fundamental discoveries to "fuel future generations of products and services."

Cisco came from the data side. Cisco was founded in 1984 by two former IT managers at Stanford to commercialize a router and companion software developed at the university. Thus Cisco is a great example of the phenomenon from the last chapter explaining why small companies appear to generate more radical innovation than large companies—the founders took the core invention from their prior employer, Stanford. In fact, they took it from their *current* employer—the founders continued to work at Stanford for 18 months after startup.

Because Cisco began with technology developed elsewhere rather than from internal R&D, it inherently followed an open innovation strategy. Thus its innovation strategy at least initially was to partner with or invest in promising startups as a way to gain new technology. Ultimately, however, Cisco spent as much on R&D as it did on A&D (acquisition and development). Thus it had moved from exclusively open innovation (using Stanford's innovation) to the more hybrid form envisioned by Chesbrough.

While the argument that the two strategies (open innovation and closed innovation) competed directly is valid, the argument that open innovation is the means for startups to unseat large established companies is debatable. Cisco was a monopolist in the router market for many years, and thus had no large companies to unseat. Cisco only had to compete with Lucent when voice and data began to merge in the late 1990s. At that point, the two firms had comparable size. Cisco's market capitalization of roughly $100 billion approached that of Lucent (roughly $150 billion).

The response to Chesbrough's prescription for open innovation has been so widespread that a recent University of California, Berkeley survey of executives from the largest companies in the United States and Europe found that 78 percent reported applying open innovation for many years. Moreover, they expected use of open innovation to grow even further.

The critical question, however, is how have companies who've adopted open innovation fared? What's the impact of open innovation on companies' RQ? Before answering that question, we first need to acknowledge that open innovation takes many forms: crowdsourcing ideas, funding research at universities and government labs, licensing technologies from those labs, fostering startup companies through corporate venture capital (CVC), joint ventures with rivals and suppliers, and outsourcing R&D. These are very diverse activities, so it's unlikely they all have the same impact. Accordingly, we review the record on two broad classes of open innovation, idea sourcing and idea development, separately.

IDEA SOURCING

One of the most prevalent forms of open innovation is idea sourcing—developing and commercializing an idea that originated outside the company. It is so prevalent that a recent innovation survey conducted by Ashish Arora, Wes Cohen, and John Walsh[2] found that 49 percent of the most important product innovations from manufacturing companies originated from an outside source. These sources included suppliers, customers, rivals, commercial consultants and labs, independent inventors, and university and government labs. Thus, there is no question that idea sourcing is widespread. The important question is one of impact—which sources yield what results?

The most common source of external ideas is customers—accounting for 27 percent of the ideas behind companies' most important innovations. This finding reinforces an important stream of research on user innovation pioneered by Eric von Hippel at MIT.[3] The principle behind user innovation is that users have acutely felt needs often not met by existing products. To satisfy these needs they could generate their own innovations. However, because they typically lack the resources to produce those innovations, the more expedient solution is to convince existing suppliers to develop them. If the user comprises a large share of the manufacturer's sales, or is representative of a class of users that comprise a large share, then it often makes sense for the manufacturer to develop that idea. A classic context for user innovation is medical devices. Companies such as Johnson & Johnson actively encourage prominent surgeons to invent devices. Not only will the inventing surgeon use the device, but knowledge that a renowned surgeon invented the device (through publications, conferences, and the manufacturer's own marketing) fuels diffusion of the device to other surgeons.

The next most common source of external ideas is suppliers, comprising 14 percent of companies' most important new products. A classic example of companies that rely on supplier innovation is Toyota. Suppliers have three advantages as a source of ideas. First, they have strong incentives to generate these ideas as a way to strengthen the buyer's dependence upon them. Second, they typically have in-depth knowledge of the customer's needs and capabilities, so can develop ideas that best match them. Finally, they have well-developed relationships with the customer that facilitate the transfer of knowledge.

One concern suppliers should have when they innovate for their buyers, however, is that the buyers provide the innovation to other suppliers. If that occurs, the innovating supplier may fail to profit from its innovation. This happened to a friend of mine, Dan Pulos, who held multiple patents for products he produced for a Fortune 50 company. That company contracted with another supplier to produce some of Dan's inventions—a clear patent infringement. The problem pursuing that claim, however, is that the Fortune 50 company accounted for approximately 50 percent of Dan's revenues. Dan couldn't afford to lose the customer, so he had to tolerate the infringement.

There do, however, appear to be strategies suppliers pursue when faced with similar threats. Jenny Kuan, Dan Snow, and Susan Helper present a rich study characterizing these for the 70 percent of automotive suppliers who contribute design work.[4] In survey responses from 1,400 suppliers, the authors found that companies employ three different strategies in dealing with these contracting hazards. The most interesting of these is a "high R&D" strategy in which the suppliers offer technologically advanced products to multiple firms. This gives them "seller power," both because they don't rely too heavily on any single buyer and because they have cutting-edge technology not available from other suppliers.

Other external idea sources are less prevalent because they lack the strong communications channels and/or the aligned incentives characterizing the customer and supplier relationships. These other sources include rivals, consultants/service providers, and independent inventors. Each comprises about 8 percent of ideas for companies' most important new products.

The least common idea source (5 percent) is universities. This is somewhat surprising and also unfortunate because university technology transfer offices (TTOs) all maintain markets for university inventions as a way to generate licensing revenue from faculty research. However, as we saw in the last chapter, the record on companies utilizing university technology is weak (university licensing revenue is only 1 percent of grant revenue). The likely reason for this is that the ideas produced by universities (other than those produced under contracts with companies) tend to be earlier stage, and therefore more remote from commercial applications.

So we have a sense of how prevalent various external idea sources are. Accordingly, we also know their relative rank. What's interesting about the relative rank is that it should reflect the relative returns. In other words, customer ideas are likely most prevalent because they generate the highest returns. This is an important insight. An interesting follow-on question, however, is whether the returns to customer ideas are high because the cost to acquire and/or commercialize them is low, or because customer ideas are better and therefore generate higher revenues.

Ashish, Wes, and John examined this as well by looking at sales revenue from new products emanating from the external idea sources. They found that a new product from customer ideas had very little impact on revenues. On average, it comprised only 17 percent of company sales. In contrast a new product from a specialist source (consultants, independent inventors, or universities) comprised 26 percent of company sales. Thus sales from specialist sources are roughly 60 percent higher than those from customer ideas even though use of customer ideas is more prevalent (27 percent versus 21 percent). This suggests the prevalence of new products based on customer ideas stems from low cost to acquire and commercialize customer ideas rather than because their quality is higher.

A newer source of external ideas that is generating considerable attention is crowdsourcing. Crowdsourcing is a means to generate ideas or services by soliciting them from a large group, typically via an online platform like Kaggle (a predictive analytics crowdsourcing site) or Topcoder (a computer programming crowdsourcing site). To create a crowdsourcing competition, the sponsoring organization characterizes the real-world problem it is trying to solve, offers a cash prize, and broadcasts an invitation to submit solutions. One such example from Karim Lakhani and Kevin Boudreau, two of the leading researchers on crowdsourced innovation, is Merck's competition on Kaggle to generate ideas for streamlining its drug discovery process. Merck hosted an eight-week contest with a $40,000 prize that generated 2,500 proposals from 238 teams. The winning solution utilized a machine-learning approach and relied upon expertise (computer science) not housed within Merck (at least at the time).[5]

This illustrates one of the advantages of crowdsourcing—it draws on a larger and more diverse set of skills, experience, and perspectives than those available within any given company. In addition, the sponsor obtains free labor. In the case of Merck, this was a sizable amount of free labor (the 237 teams and 2,499 proposals that didn't win the $40,000 prize). If we assume that each team consists of three people working one-fourth time (10 hours per week) for the eight-week period, and we further assume the national average salary of roughly $100,000 (before benefits) per developer, Merck obtained $28.6 million of labor for $40,000!

Finally, the sponsor obtains free experimental results from the teams and proposals that didn't win. These nonwinning proposals provided Merck a much deeper sense of what the solution space looked like—including what solutions didn't work and why. Moreover, while the knowledge is crowdsourced, the sponsor maintains all the intellectual property for both the successful and unsuccessful solutions.

One question that remains underexamined with respect to crowdsourcing ideas is the extent to which they are implemented by the sponsoring organization. While the crowdsourced solution to the screening problem was ultimately implemented at Merck, there is evidence suggesting this is the exception rather than the norm. Implementing such solutions seems to fall victim to classic knowledge transfer problems.

Successful transfer of external knowledge requires a number of stars to align. In particular, the external source and internal receiver both have to be motivated to make the exchange, the knowledge has to be inherently transferable, and the recipient has to have sufficient existing knowledge to absorb the new knowledge. Finally, the receiver needs to feel the knowledge is valid. Often, however, external knowledge is automatically dismissed as being inferior—the classic Not Invented Here (NIH) syndrome. This may be particularly acute in the case of crowdsourcing since the source is anonymous. Indeed, a field study of a startup company[6] (with 50 scientists) ran four crowd contests over a three- to six-month period. The study found that while some prizes were awarded, none of the ideas were used in subsequent R&D efforts. Interviews with the companies' employees indicated that the failure to implement the solutions stemmed from perceived quality. The company's scientists felt the solutions were not relevant (did not meet all the contest criteria), and for those that were relevant, they felt the solutions tended to resemble ones the company had already considered. Note of course this could be fact, or could merely reflect NIH syndrome. Regardless, it suggests crowdsourced ideas may pose implementation problems.

Thus the two big questions about the long-run viability of crowdsourcing are (1) whether highly skilled participants will continue to offer free labor, and (2) the extent to which the crowdsourced ideas are implemented by the sponsoring company.

IDEA DEVELOPMENT

While *idea sourcing* considers open versus closed innovation at the front end of the R&D pipeline, *idea development* pertains to open versus closed innovation at the middle of the pipeline. Open innovation at the development stage can take many forms such as joint ventures, collaborations, and so on. However, we focus attention on the most extreme form of open innovation at this stage: outsourced R&D. This focus stems from the fact that the NSF has data on outsourced R&D over a longer period.

What is most striking about the NSF data is that it indicates outsourced R&D increased by a factor of 20.5 (2,050 percent) during a period in which R&D itself (measured as the number of scientists and engineers working within companies) increased by roughly one-eighth that amount (250 percent). Of course, the real question regarding outsourcing R&D is not the prevalence, but rather the effectiveness.

Not surprisingly companies believe their shift toward outsourced R&D has benefited them. In fact, a survey of CIOs and CEOs revealed that 70 percent of them believe outsourced innovation improved their financial performance.[7] However, we know from Chapter 1 that companies have had no good way to gauge whether this is true. So to truly understand the effectiveness of outsourced R&D we need to examine its impact on RQ.

To do that, I separated companies' R&D into two buckets—R&D conducted internally, and outsourced R&D—and

treated them as separate inputs for producing company sales. Doing this allowed me to determine how productive each form of R&D was—how much a 10 percent increase in each of internal and outsourced R&D increased company revenues. When I looked at internal R&D, I found that on average, a 10 percent increase in R&D spending increased revenues 1.3 percent (while this sounds like the company is losing ground, revenues are typically 50 times larger than R&D expenditures).

However, when I looked next at outsourced R&D, I obtained the astounding result that *outsourced R&D had an RQ of zero*! This means a 10 percent increase in outsourced R&D yields *no increase* in company revenues. As a result, the company's profits actually *decrease* by the amount of spending on outsourced R&D.

This result was almost implausible, so I did additional analyses to understand what might be going on. The first analysis I conducted was interviewing companies about when and why they outsource their R&D. These interviews indicated that companies outsource R&D for a number of reasons, and that those reasons vary with top-level R&D strategy rather than industry mandate (in less than 5 percent of industries is it the case that all companies outsource).

At one end of the outsourcing spectrum, companies outsource only to universities and government labs. They do this to gain access to basic research as well as to identify potential employees. In the middle of the spectrum, companies outsource under special circumstances. For example, they use outsourcing as a flexible substitute for internal hiring when future demand is uncertain; they outsource activities where they lack capability and don't intend to develop it internally (because they would operate below efficient scale); or they outsource testing (particularly in the case of pharmaceutical trials by contract research organizations, CROs). At the extreme end, companies outsource all "noncore" R&D activities.

Unfortunately, the NSF Survey of Industrial Research and Development (SIRD) did not collect data on the destination for outsourced R&D; however, these data were collected in its successor, the Business R&D and Innovation Survey (BRDIS). A report of BRDIS data[8] indicates 3.4 percent of outsourcing is to universities, 81.3 percent is to companies, and the remaining 15.2 percent is to government agencies and other organizations. Since the vast majority of outsourcing is to companies, outsourcing seems to conform to rationale in the middle and extreme ends of the spectrum.

The second analysis investigated how reliable the result of unproductive outsourced R&D actually was. To examine that, I looked at what was happening to the productivity of internal versus outsourced R&D over time. I found that there had been no change in the productivity of either internal R&D or outsourced R&D. This meant outsourced R&D had *never* been productive. This was a reassuring result, because it meant I didn't have to worry about how outsourced R&D might have been changing.

The third analysis examined whether outsourcing was truly unproductive, or whether it just appears that way because poor quality companies are the ones that outsource their R&D. This turned out not to be the case. The internal RQ of outsourcing companies is the same as that for companies who don't outsource at all.

The fourth analysis checked whether outsourced R&D just appears to be nonproductive because companies outsource their worst projects. This idea came from discussions with a colleague who works with Merck. He said that Merck outsources the projects it thinks it might want to kill because it is easier to kill outsourced projects than it is to kill internal projects. To kill an outsourced project, you merely fail to renew the follow-on contract. In contrast, killing an internal project requires fighting political battles that may alienate the companies' researchers.

To check whether companies were outsourcing their worst projects, I looked at what happens when they first start outsourcing their R&D. If they truly outsource their worst projects, you would expect their internal RQ to go up at that point. This is because getting rid of the bad projects leaves the company with a stronger portfolio of projects. This stronger portfolio should generate higher returns (RQ). That's not what happens. In fact, internal RQ is slightly higher before outsourcing, but by an amount so small it's not worth considering. Thus there is no evidence project quality is driving the lower productivity of outsourcing.

In short, after checking which companies outsource, who they outsource to, and what they outsource, it appears outsourced R&D is inherently less productive than internal R&D. Why might that be? While the data don't provide insight, one possible explanation is that R&D produces internal spillovers—knowledge that gets generated by one project but can be utilized in other current or subsequent projects as well. To understand the intuition for internal spillovers, remember that companies carry portfolios of projects. As we know from Chapter 2, the most likely outcome from these projects is termination (the 123 of 125 projects that are never launched commercially). This means the value of R&D projects likely lies outside project outcomes themselves—possibly in the ability to recycle knowledge gleaned from the failures into subsequent projects.

I provided one example of technology reuse in Chapter 2—the redeployment of ion beam technology at Hughes Aircraft

Company from satellites to semiconductors. Had Hughes outsourced the ion beam R&D, the outsourced company would have derived the benefit from any other application (to the extent that they too were broad-based). While the Hughes case is anecdotal, there is some quantitative evidence that companies fail to capture spillovers when they outsource. A recent study by Carmen Weigelt of Internet banking adoption finds that banks that outsource their initial IT integration are less able to develop new applications and accordingly have lower revenues from their Internet operations.[9]

An alternative explanation for the lower productivity of outsourced R&D came from Nick Heinz, chief financial officer of Sears Home Services, who used to share my St. Louis to Los Angeles commute. On one of those flights, I told Nick about the finding that outsourced R&D was unproductive and said I was trying to understand why. He said, "Oh, I know why. It's the 'consultant effect.'" He explained that only 30 percent of consulting recommendations are adopted, because all the knowledge to implement them resides with the consulting company, since it did all the investigation. As a result, the funding company would have to replicate much of the work it had already paid for to have sufficient expertise to implement the consultant's recommendation. This explanation is reminiscent of the problems discussed earlier regarding implementation of crowdsourced ideas.

Perhaps the simplest explanation of why outsourced R&D is unproductive is that the resources at the supplier company are inferior. This would explain both why companies outsource R&D and why outsourced R&D is less productive. Quite simply, for the supplier company to offer a contract that provides it a profit, while still being less expensive than conducting the project internally, the supplier has to have lower cost inputs.

This point is made vividly in Pierre Azoulay's study on pharmaceutical manufacturers outsourcing their clinical trials to clinical research organizations (CROs). One outsourcing manager in the study complained, "CROs keep giving us bad people to choose from, and there is nothing I can do about it," and another complained that the CRO implied that an "A-Team" would be conducting the trial, but "rookies" had been substituted at the last minute.[10]

Beyond the questions of why outsourced R&D is unproductive, a final question I'm typically asked when presenting these results is why companies persist with R&D outsourcing, given its lack of productivity. Again the NSF data provide no insights. However, I believe it's because companies don't yet realize that outsourcing is unproductive. This would be an obvious extension of the problem introduced in Chapter 1 that companies lack good measures of R&D productivity.

Given this ambiguity about the productivity of R&D, companies are vulnerable to "information cascades." An information cascade occurs when companies discount their own instincts and follow the behavior of other companies or recommendations of consultants who they believe have superior information.[11] Given the attention to open innovation, and the 20.5-fold increase in outsourced R&D, it's likely companies believe outsourced R&D is productive, even though there is no evidence other than widespread adoption.

THE BIGGER PROBLEM

Unfortunately, the profit decrease from outsourced R&D is only the tip of the iceberg. Outsourced R&D is a slippery slope. Once a company begins to outsource R&D it previously conducted internally, it closes labs and reduces technical staff. As a result, the next time a company has a similar project that it considers conducting internally versus through outsourcing, the deck is stacked in favor of outsourcing. That's because executing the project internally requires rebuilding the labs and hiring new technical staff, in addition to the marginal cost (labor, equipment, and materials) of the project itself. In contrast, outsourcing just requires the marginal cost of the project. Because each new outsource decision has this flavor, ultimately the company's R&D capability unravels and its RQ decays.

This unraveling of R&D capability appears to have happened at a major division of a Fortune 50 company. Initially the division began outsourcing drafting to India as a way to satisfy its trading offset requirements. Offsets are a common condition of international trade contracts that requires the foreign exporter to purchase goods or services from the importing country. Drafting seemed to be an innocuous form of outsourcing because it doesn't involve scientists or engineers. However, the next step in the unraveling was that the division outsourced design, rationalizing that the heart of the design was the standards (which it retained in-house). Ultimately, however, if you don't do design, you lose the requisite knowledge to determine whether designs meet standards, so you can't develop standards either. In the end, the division was forced to outsource standards writing as well. Thus outsourcing of R&D not only has zero RQ itself, it leads to further outsourcing, and ultimately to the inability to conduct R&D internally. This unraveling of R&D capability as a consequence of outsourcing R&D is related to the unraveling of innovative capability as a consequence of outsourcing manufacturing that Gary Pisano and Willy Shih document in *Producing Prosperity*.[12]

SUMMARY

There is a widely held belief that open innovation increases companies' financial performance. Accordingly, open innovation has been adopted by the vast majority of companies engaged in R&D. More telling, there has been a 2,050 percent increase in the amount of outsourced R&D.

While there is some evidence that open innovation in the form of idea sourcing may improve companies' financial performance, the record on idea development indicates that R&D outsourcing not only fails to improve financial performance, it actually degrades it! This occurs because outsourced R&D incurs R&D expenditures without increasing revenues. Thus it decreases profits. Worse, however, it appears that outsourcing R&D is a slippery slope wherein company innovative capability decays, so the company increasingly outsources, and capability decays even further.

Overall, these mechanics are so powerful that outsourced R&D accounts for much of the 65 percent decline in RQ that we saw in Chapter 1. The RQ for internal R&D has remained relatively constant, which is great news! It means if companies are willing to undertake the investments to recreate labs and rebuild their technical staffs, over time they should be able to restore RQ to prior levels.

MISCONCEPTION 6: R&D Needs to Be More Relevant

Another major trend in R&D over the past few decades has been tighter coupling with marketing—making R&D more commercially relevant. This has led to a shift in the locus of R&D from central corporate laboratories to operating divisions' laboratories, and from centralized decision-making to decentralized decision-making. Of all the prescriptions discussed so far, the need for more relevance is the one most widely viewed to be valid. In an informal survey of professionals attending RQ workshops, 80 percent of consultants and 90 percent of investment analysts and managers believe that having R&D decisions made by divisions (which are closer to the commercial markets) leads to higher RQ.

THE IMPETUS FOR DECENTRALIZATION

This movement toward decentralized R&D began during what Eliezer Geisler calls the "Renewal Period" of industrial science and technology (1984–1993).[1] Two trends in this period had considerable impact on companies' R&D. First, President Ronald Reagan sharply increased the Department of Defense (DoD) budget—much of which comprises industrial R&D. Second, U.S. company dominance in a number of industries, such as steel and automobiles, was threatened by foreign competition. The former provided increased funding for R&D; the latter provided the motivation to innovate.

The first trend, increased defense funding, tended to reinforce centralized R&D. This is because DoD projects are often aimed at pushing the technology frontier to compete in the arms race, which typically involves basic and applied research. However, the second trend, threat from foreign competition, pushed companies in the opposite direction. The principal company response to the foreign threat was to implement total quality management (TQM) programs. The role of R&D in these programs was to generate process improvements to reduce unit cost. This form of R&D is inherently product-specific—redesigning the product to make it more manufacturable, or redesigning the manufacturing process itself. Thus Geisler attributes the shift toward highly relevant, short-term, and product-oriented R&D directly to the mobilization of R&D to support TQM reengineering programs.

This shift toward product-specific R&D reached maturity in the subsequent period and continues today. Accompanying the shift has been a radical decrease in basic research, from roughly 10 percent of companies' R&D prior to 1984 to approximately 3 percent in the 2000s, as well as movement from central laboratories to divisional laboratories. As central labs began to dissolve, divisions gained more control over the funding of R&D. They largely preferred relevant research with short-term results and low risk.

A concrete example of this phenomenon comes from the experience of Hughes Research Laboratories (HRL) during this period.[2] HRL was formed in 1954 as the third division of Hughes Aircraft Company (Radar and Guided Missiles being the other two divisions). We saw Hughes earlier in the ion beam technology story (Chapter 2) and the laser story (Chapter 5). The province for HRL was applied research—coming up with new components and technology that would further the business interests of Hughes. The approach to fulfilling this goal was to hire scientists and engineers with expertise in the technologies related to Hughes' product lines and to give them free rein. Though there was no grand strategy for the lab, it grew from a population of 200 from the time it moved into its quarters in Malibu, California (1960), to a population of nearly 560 at its peak in 1987.

The forces affecting Hughes during the "Renewal Period" were different from those just discussed, in large part because defense companies typically don't face threats from foreign competition. The more relevant forces to Hughes were anti–defense industry sentiment that arose in response to Reagan's increased defense spending, the end of the cold war, and the 1985 sale of the company to General Motors. Interestingly, despite the difference in stimuli, the Hughes response mimicked that of other industrial companies during the period.

One such response in 1988 was the direction from then-president Malcolm Currie for HRL to give the sectors (divisions) "more wheels." Up until that time very little (less than 5 percent) of HRL's internal funding came directly from sectors. This occurred when work at the lab was necessary to support sector contracts or internal R&D projects.

In general, it was felt there was little need for sector funds. First, the sectors each had their own R&D efforts to support their forecasted programs, and second, the time horizons of HRL and the sectors differed. HRL was interested in technologies that weren't expected to have benefit for 5 to 10 years, while the sectors were primarily interested in technologies for known programs. However, when the sectors began downsizing in response to the various environmental elements, they began to challenge corporate to make HRL more relevant: "If I have to cut R&D costs and undergo downsizing, I would like HRL to be more relevant to my needs and more near-term directed."[3]

One factor facilitating a transformation of HRL toward "giving the sectors more wheels" was a $10 million shortfall in the construction costs for a second HRL building. Dr. Art Chester, the HRL director, exploited the shortfall to develop a proposal recommending (1) that corporate cover the shortfall, but (2) that each sector be given say over $1 million of the research budget that had already been approved for HRL. If implemented, the proposal would increase sector-directed work from 5 percent to 20 percent of HRL funding. While that was already quadrupling sector funding, the new company president, C. Michael Armstrong, whom GM hired from IBM, established an even higher goal of 50 percent sector funding.

The issue of greater sector funding was hotly debated both at HRL and the sectors. One HRL lab manager felt having directed money that could not be countermanded was important. It provided a clear statement of customer (sector) objectives. However, she also felt it was important that HRL continue to have discretionary funds to shape the direction of future technology. Thus in her view the issue was more one of balance than the appropriateness of sector funding.

Others at HRL, however, felt that the idea of allowing sectors to direct HRL work was inappropriate. One principal scientist felt that HRL should be advising sectors on what was important rather than the other way around—that to be a world-class lab, HRL needed to be looking further ahead than sectors: "Sectors can do their own short-term work better and more cheaply than we can."

Even at sectors, there were those who felt that the move toward greater sector funding of HRL might jeopardize its contributions. Jim Bailey, a technology manager at one of the sectors, felt that HRL provided two valuable services: it kept tabs on broader technology, and it advanced new technologies. He felt HRL should be doing basic research broadly applicable to Hughes's businesses. To illustrate his concern, he referred to one lab manager at HRL who had not gotten enough of his tasks funded by sectors. The lab manager was given undirected money for those tasks for one year but was told it was not going to happen every year—"If there is no sector interest, maybe HRL shouldn't be doing these tasks." Bailey commented that "maybe HRL *should* be doing those tasks. Maybe they should be doing *exactly* those things no one at the sectors knows are needed."

We can't tell the outcome of Hughes's decentralization on RQ because General Motors later affected even greater decentralization by selling Hughes off in pieces: the military businesses for $9.5 billion to Raytheon in 1997, the satellite business for $3.75 billion to Boeing in 2000, and the DirecTV business for $26 billion to EchoStar in 2001.

However, we can see the impact of other companies decentralizing their R&D. Procter and Gamble (P&G) began decentralizing R&D as part of Durk Jager's (CEO from

January 1999 to June 2000) "Organization 2005" initiative. The goals for the initiative were to grow global revenues from $38 billion to $70 billion, while also raising profitability. The decentralization of R&D was intended to accelerate innovation—in particular, time to market.

Jager felt decentralization of R&D was necessary to make the big company "feel small" (provide greater managerial control, break bureaucratic inertia, and break ties to current technology), as well as to more closely fit the needs of divisions and markets. Note that while projects themselves would be decentralized, "back office" functions would remain centralized to achieve economies of scale. While Jager was the main force behind decentralization of R&D, A. J. Lafley, who replaced Jager after his resignation in June 2000, maintained a commitment to it.

The result was a dramatic shift in P&G from 90 percent centralized control of R&D resources in the 1990s to 90 percent decentralized control of R&D in 2008. Prior to the decentralization, P&G was known for creating entirely new product categories: first synthetic detergent (Dreft in 1933), first fluoride toothpaste (Crest in 1955), and, more recently, Febreze odor fresheners (1998), Swiffer (1999), and Crest Whitestrips (2001). (While Crest Whitestrips were introduced after decentralization, all their development took place prior to that time.)

In the words of R&D chief Bruce Brown, making business-unit heads responsible for developing new items inadvertently slowed innovation by more closely tying research spending to immediate profit concerns.[4] Relatedly, it led to smaller, more incremental innovation. While the number of innovations doubled, the revenue per innovation decreased 50 percent.[5] P&G has failed to introduce a single blockbuster since the decentralization. This is likely because the big blockbusters in the past came from generating new technologies, or leveraging technologies across the business units. Crest Whitestrips, for example, combined bleaching technology from the laundry business, film technology from the food wrap business, and glue technology from the paper business. Once these businesses were decentralized, there was no convenient mechanism for linking technologies in this fashion.

A large part of the decline in P&G's innovation could be because R&D intensity declined 44 percent following decentralization (from 4.3 percent of sales in 2000 to 2.4 percent of sales in 2011). While a number of other events occurred during that period (such as the Gillette acquisition and the economic downturn), some of this spending decline is likely a direct consequence of decentralization. When business units control their own R&D budgets and are compensated for short-term results, they typically cut R&D to pursue investments with shorter-term payoff. The consequence of cutting R&D at P&G was that the company seemed to have cut meat as well as fat. Its RQ fell from 103 in 1999 to 90 in 2007 (while not obvious because RQ is rescaled to match the IQ scale, this represents over a 50 percent decrease in R&D productivity).

The encouraging news is that P&G's RQ has since returned to 100. Much of this return can be attributed to efforts to recentralize some of the R&D activity and to better manage the tension between centralization and decentralization. In 2004 P&G initiated efforts to create a new growth factory based on principles from disruptive innovation. The charter of the factory was to generate new growth initiatives of three types: create new product categories (such as Tide Dry Cleaner franchises), bolster existing product lines (such as Crest 3D White), and deliver major new benefits in existing product categories (such as the Gillette Guard, a low-cost, single blade razor created for the Indian market). In addition, P&G revamped its strategy development and review process to integrate company, business, and innovation strategies.[6]

CENTRALIZATION VERSUS DECENTRALIZATION

The theory underpinning centralization versus decentralization of R&D comes from the broader theory of company centralization or decentralization. The tension between the two forms receives considerable attention in the academic literature. The most provocative treatment is Jackson Nickerson and Todd Zenger's theory that companies vacillate between the two forms to obtain the benefits of both because there is no organization design with the appropriate balance between them.[7]

The advantages of decentralization resemble those of small companies that we discussed in Chapter 2: improved information processing and better ability to attract and drive performance from high-quality employees. In contrast, the advantages of centralization resemble those of large companies, such as scale economies. However, what's unique about the centralization/decentralization debate in the R&D context is that centralization provides opportunities to conduct research that spans the organization, is longer term, and is broader in scope. In contrast, decentralization leads to parochial innovation that tends to have a shorter-term perspective.

The Record on Decentralized R&D

We've seen the forces pushing companies in the direction of decentralized R&D, we've heard views on both sides of the debate, and we've seen two examples of decentralization leading to lower innovativeness. But what does the broader record tell us? This is a difficult question to answer because there is no readily available data on company decentralization. So we're going to tackle this question using two approaches to measuring decentralization.

The first approach uses data from a survey of R&D executives that the Industrial Research Institute (IRI) conducted in 1994, and whose results were published in 1995. The degree of R&D centralization in the survey was measured on a five-point scale: 1 = decentralized; 2 = decentralized hybrid; 3 = balanced hybrid; 4 = centralized hybrid; 5 = centralized.

Nick Argyres and Brian Silverman, who were the first scholars to examine the impact of centralization on company innovation, used these IRI data in their analysis. In order to link "decentralization" to company performance, they needed to match the IRI data to publicly available financial data as well as patent data. This reduced the total number of companies they could evaluate in half. Once they had done this matching, they compared a company's level of centralization to its patent citations. They found that centralized R&D produces innovations that have a larger and broader impact on subsequent technological evolution. Thus their results seem consistent with the experience at P&G.[8]

Nick and Brian were fundamentally interested in the impact of centralization on the importance of company innovation (how influential the research was in driving follow-on research). I was interested in the related, but separate, question of centralization's impact on the productivity of innovation (RQ). Accordingly, they were generous in sharing their data with me. The results of my analysis comparing the IRI centralization measure to RQ are presented in Figure 7-1. They are consistent

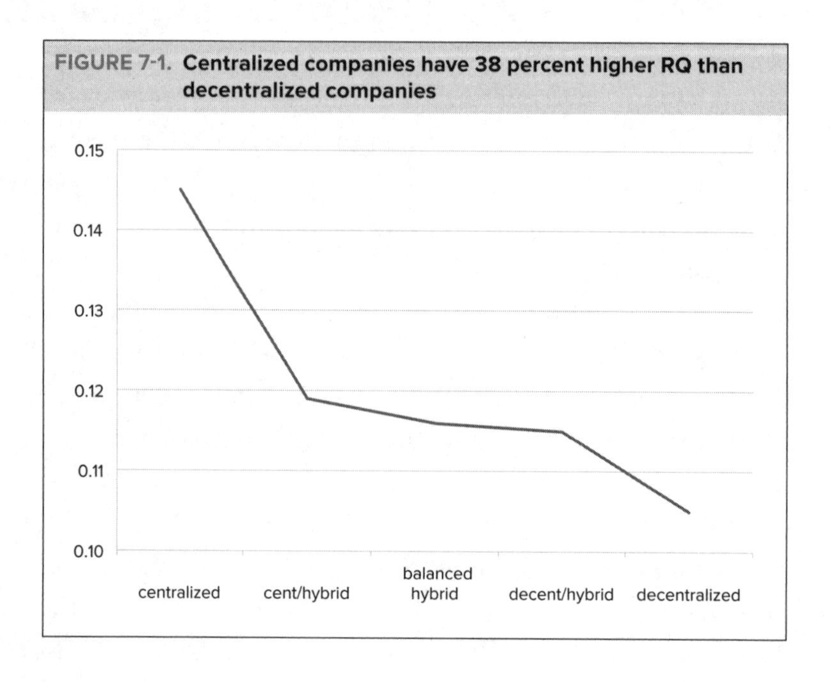

FIGURE 7-1. **Centralized companies have 38 percent higher RQ than decentralized companies**

with Nick and Brian's result as well as the outcomes following decentralization at HRL and P&G. They indicate that companies with centralized R&D have 38 percent higher RQ than companies with decentralized R&D.

This first approach to evaluating the impact of centralization on RQ uses a reliable measure of centralization, but it is based on a very small set of companies. So to complement those results, I also examined an additional centralization measure that allows us to look at a broader set of companies. This second measure comes from work by Ashish Arora, Sharon Belenzon, and Luis Rios, all from Duke University.[9] The Duke measure computes the level of R&D centralization as the percentage of a company's patents assigned to the company's name (as opposed to being assigned to one of the company's affiliates). A "decentralized company" is one whose value for this measure is in the bottom fifth of the sample of 1,014 companies; a "centralized company" is one whose value is in the top fifth. While the measure is very clever, an important caution is that typically an operating division has the same name as headquarters unless it has been acquired. Thus this measure may be capturing the impact of acquisitions on RQ, rather than the impact of decentralization on RQ.

With this caveat in mind, the results are similar to the prior results for the smaller sample with the better centralization measure. In both cases, the average RQ for centralized companies is substantially higher than that for decentralized companies. In fact, the difference is more dramatic using the patent definition. Centralized companies have 64 percent higher RQ than decentralized companies when using the patent-based measure; centralized companies have only 38 percent higher RQ when we use the survey measure of centralization. This difference in results likely stems from the fact that subsidiaries are typically managed autonomously, thus losing all the benefits of scale and knowledge sharing from the broader company.

Why Centralized R&D Yields Better Results

The logic underlying the push for decentralization and greater relevance of R&D is that R&D directed by divisions will be more responsive to what the market wants. Accordingly, it is more likely to yield a financial payoff. This is certainly true (as is the fact that the payoff is more quantifiable and less risky—features favored by financial managers). However, the vulnerability in that logic is captured in the Steve Jobs quote, "A lot of times, people don't know what they want until you show it to them."[10] Thus the first problem with relevance is that its key assumption (that the market offers the best information regarding what R&D to engage in) may be wrong. Indeed, we saw in the last chapter that ideas that come from customers have the lowest value—they yield the lowest level of new company sales. Once we know that customer ideas yield the lowest level of new sales, it is not surprising that decentralizing R&D to be more responsive to customers will also lead to lower RQ.

The more pernicious problem with decentralization, however, is that when R&D is directed by operating units, the company fails to produce the early stage technology that opens up new opportunity. This occurs because operating division managers tend to be evaluated on their divisions' profitability, rather than the company's market value (which is the basis for most CEO compensation). The distinction between current profits and market value is important because market value takes into account future profits, so in principle it captures the long-run returns to R&D. In contrast, current profits penalize R&D, because accounting rules require R&D to be expensed. This means all R&D is subtracted from operating income in the year it is expended, while the payoffs of R&D don't occur until future periods. Thus the further out the fruits of R&D, the less likely operating divisions are to conduct it.

A final problem with decentralizing R&D is that operating divisions have little or no incentive to conduct R&D that benefits multiple divisions, since there typically aren't mechanisms to share the costs of that R&D in a decentralized system. Relatedly, even if divisions do conduct research that can be utilized by other divisions, without a centralized R&D organization there isn't (1) a mechanism for knowing which (if any) other divisions can utilize the knowledge, nor (2) a mechanism for diffusing the knowledge even if it's known that another division can utilize it.

The combined effect of having a nearer-term orientation as well as a more parochial orientation is that decentralized companies conduct less basic research. One implication of conducting less basic research is that companies are less likely to generate inventions that open up new opportunities. An additional implication of conducting less basic research is that it increases the likelihood of a firm outsourcing R&D. We know from the last chapter that conducting R&D internally versus outsourcing has a dramatic impact on RQ.

This shift away from basic research can happen even if there is a central lab, provided the funding for that lab is controlled by divisions. Ultimately the pressure for relevance from the divisions may yield a central lab that is incapable of generating innovations, even when divisions know what the market wants.

A concrete example of this comes from a Fortune 1000 manufacturer of precision mechanical components. In the past, the company had a large R&D budget controlled by a centralized lab that "figured out the right things to do with it"—a charter very similar to Hughes Research Laboratories. Over the past 20 years there has been a 180-degree shift in the orientation of the company's budget such that the operating divisions drive R&D. At one point after the shift, the central lab proposed developing technology to improve a key dimension of performance across the company's product lines. The operating divisions determined the technology wasn't a priority, so development was cancelled. Unfortunately, a few years later, competitors introduced the technology in their products and it killed demand for the company's own products, forcing it to lower prices. At that point, the operating divisions came back to the lab wanting the technology right away. However, because it had never been funded, it still required two to three years of development. Since the divisions couldn't afford the lag, the company was forced to outsource the technology.

SUMMARY

A number of economic forces in the 1980s and 1990s led to widespread decentralization of R&D. The logic of decentralized R&D is that it makes companies more responsive to the market. The view that this is beneficial is widely held—in an informal survey, approximately 80 percent of consultants and 90 percent of investment professionals believed decentralized R&D is associated with higher RQ. In fact, the opposite is true: companies with centralized R&D have 40 percent to 64 percent higher RQ than companies with decentralized R&D. This occurs because companies with centralized R&D tend to (a) do more basic research, so are more likely to create new technical possibilities, (b) create technology that benefits multiple divisions, and (c) derive more of their technology from internal R&D rather than through outsourcing.

MISCONCEPTION 7: Wall Street Rewards Innovation

We learned in Chapter 4 that RQ allows companies to determine whether to increase or decrease R&D investment, and by how much. It also allows them to quantify the expected impact of the change in R&D on sales, profits, and market value. Given this ammunition, a request to increase the R&D budget in underinvesting firms should be a slam dunk for CEO approval. However, when I've proposed this to companies in the past, the typical reaction is, "There is no way we can increase R&D that much—the shareholders won't allow it."

Initially I was surprised by this reaction: (1) how could shareholders wield so much power, and (2) why would shareholders prevent companies from doing something in their own best interest? Then I realized it's because shareholders don't understand what's in their best interest—they are subject to the same "flying blind" problem confronting companies.

As a crowning example of that, Warren Buffet, perhaps the most sophisticated investor of all time, has historically eschewed technology stocks, precisely because he didn't have a way to value them. He once joked to a class that he would fail any student who gave an answer to an exam question asking them to value an Internet company.

HOW *DO* INVESTORS VALUE INNOVATION?

How do investors value innovation currently? Trade Radar (one of the most followed authors on Seeking Alpha, a crowdsourced content service for financial markets that *Kiplinger's* named as Best Investment Informant) identifies seven factors to consider with tech stocks.[1] Of these, all but one are factors to consider for any stock: (1) year-over-year comparisons versus sequential quarterly results, (2) gross margin, (3) average selling price (ASP), (4) debt, (5) customer base, (6) license (subscription) revenue, and (7) R&D. The only factor specific to tech was R&D. And here the only advice was to beware when R&D expenses appear to be in steady decline. "This could be a sign that they are milking current product lines while ignoring the future." Trade Radar is certainly right that R&D decline is something you want to investigate (though as we saw in Chapter 4, many public companies do need to decrease their R&D). However, watching R&D decline doesn't help you value innovation; it merely tracks the input to innovation. So these seven factors won't solve Warren Buffet's problem.

A more sophisticated approach to valuing innovation is Holt's Innovation Premium (IP). Holt is a division of Credit Suisse that develops tools for evaluating investments. Its IP measure is the basis for Forbes's annual innovation ranking. IP is defined as "the difference between market capitalization and a net present value of cash flows from existing businesses. [This] is the bonus given by equity investors on the educated hunch that the company will continue to come up with profitable new growth."[2] In this sense, IP is similar to an academic measure called Tobin's Q, developed by Nobel economist James Tobin, that compares the market value of the company to its total assets. While IP goes beyond Trade Radar's seven factors, it is not fundamental analysis that allows you to value innovation. It is almost the opposite—it reports back to investors how much value they have *already* attached to a company's innovation. Accordingly, IP does not correlate with subsequent returns, as noted on the Forbes website: "(IP) is also not a statement about expected excess returns—in fact . . . we went back through the data for the past 20 years and find that there is . . . no correlation of IP with subsequent return to investors."

Not only is IP uncorrelated with future returns, it is also uncorrelated with RQ. Figure 8-1 plots the RQ and IP values for companies in the Forbes 100 (the public companies with the 100 highest IP values).[3] Each diamond in the figure represents a company. To locate a company's IP, merely drop a vertical line from the diamond to the x-axis; to locate its corresponding RQ, draw a horizontal line to the y-axis. You can see from the figure that more than half of the Forbes 100 companies have below-average values of RQ (the RQ scale mimics the IQ scale, so the average is 100). Thus IP appears to be capturing companies whose growth is coming from something other than R&D. Accordingly that growth may not be sustainable.

The approach to valuing R&D that best conforms to fundamental analysis comes from Aswath Damodaran, a finance professor at NYU.[4] Professor Damodaran notes that the biggest

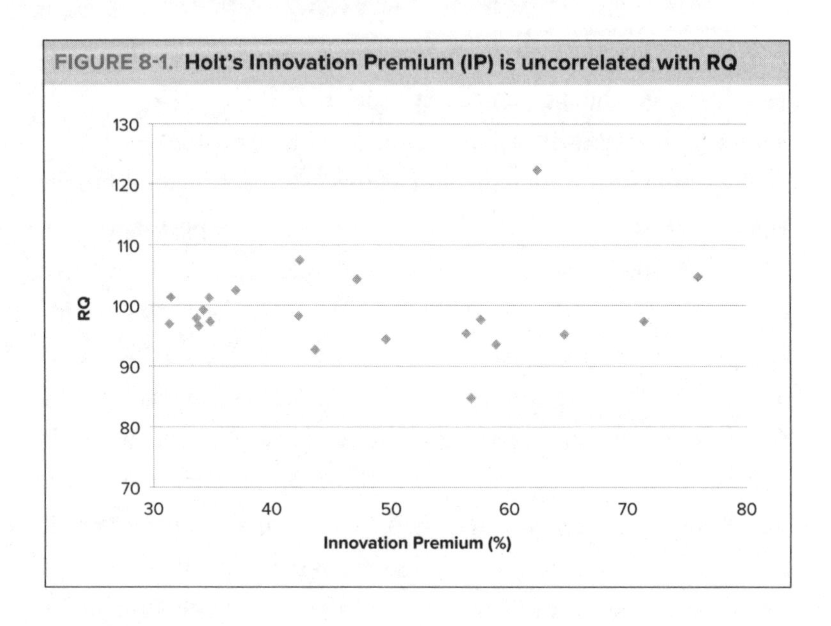

problem with treating R&D as an operating expense rather than as a capital expense (as required by the Financial Accounting and Standards Board [FASB])[5] is that investors lose the most potent tool for both estimating growth and checking for internal consistency. This is because the resulting measures of company reinvestment rate and return on capital are meaningless when the biggest asset is off the books.

Damodaran therefore offers a valuation approach that attempts to capitalize R&D. This approach requires (1) estimating the amortizable life of R&D, (2) collecting, then amortizing, all years of R&D expense across the years of life, (3) adjusting each year's operating earnings to add back that year's R&D and subtract the amortized R&D for all the other years, (4) adding the unamortized portion of R&D for all years to both the book values of both equity and capital, and (5) using those book values to determine adjusted values for return on equity and return on capital.

While this approach certainly solves the problem of uncapitalized R&D, it implicitly treats R&D and physical capital as equivalent. While it would be fairly straightforward to adjust the analysis for R&D and capital to have separate returns, the role of R&D is to increase the productivity of physical capital over time. Thus R&D should be an accelerator of the returns to physical capital.

How Do Investors Behave When They Lack Means to Value R&D?

We see that investors, like companies, have lacked the means to value R&D. How does that affect their behavior? We saw that when *companies* lack measures of R&D effectiveness, their R&D investment is suboptimal, and their RQ tends to decline over time. *Investors* don't have those problems, but they have an equivalent set of problems that hurt them as well as the companies they invest in. The first problem is that investors have to rely on faith that R&D should generate growth. Faith is hard to maintain, particularly when the alternative is to cut R&D. Cutting R&D immediately increases profits—it reduces expenses, with no detrimental impact on current revenues.

This problem of faith is captured in the sentiments of Shannon O'Callaghan, an analyst at Nomura Securities International, who covers 3M:

> 3M clearly has more of a long-term focus [than many publicly traded companies] and less desire to please investors every quarter. . . . That's admirable up to a point, but when do you acknowledge the expectations of the equity investors and finally give them what they've been looking for?[6]

Note that O'Callaghan holds these sentiments even though 3M has kept operating margins above 20 percent and has increased its dividend in each of the past 55 years.

More powerful than analysts who pressure companies for current profits are activist investors such as Nelson Peltz of Trian Fund. Rather than merely discouraging R&D investment, Peltz pressured DuPont to split into two divisions, arguing that the promise of combining chemical R&D and biological R&D wasn't bearing fruit. Ultimately Peltz prevailed, though through a circuitous route of DuPont first merging with Dow, then splitting the merged entity into three companies. In anticipation of the merger, DuPont dismissed 50 percent of its scientists at the central research lab and cut its 2016 R&D budget roughly 10 percent relative to the $1.9 billion it invested in 2015.

Without knowing DuPont's RQ, it was almost impossible for then CEO Ellen Kullman to fend off Peltz. There was no way to quantify the returns to DuPont's R&D, nor a way to demonstrate whether the company was overinvesting or underinvesting. The best defense the company could raise appeared to be a 2014 report that estimated that DuPont's combined know-how in chemical and biological engineering generated about $400 million in annual revenues (approximately 1 percent of total revenues). That analysis in itself seemed to support Peltz's case, since DuPont was investing approximately 6 percent of revenues in R&D each year. Thus a stronger defense would be to aggregate the R&D for all projects and show the increase in revenues expected from each of them. This is precisely what RQ does without having to get into the weeds of itemizing all the projects and matching them to revenues.

Because we have RQ, we can now rationally analyze whether Nelson Peltz was right about DuPont overinvesting in R&D. The unfortunate answer (because it's always tragic to see a central lab dissolve) appears to be yes. Given DuPont's RQ, a 10 percent increase in R&D ($200 million) only generates a 1 percent increase in gross margin ($156 million). Thus DuPont would increase profits by decreasing its R&D.

Activist investors are relatively rare, so it's useful to think about another important investor group that wields significant power over companies, institutional investors. In general, institutional investors are assumed to have a longer-term perspective than other investors. Focusing on the long term rather than current period profits should favor R&D, but what does the record show? A recent study by Trey Cummings at Washington University shows that companies with institutional investors have higher RQ and invest more in R&D. This leads naturally to the chicken-and-egg question of whether institutional investors make companies more innovative (increase their RQ) or are merely better at identifying innovative companies (those with higher RQ). Trey's results indicate that companies become more innovative when the number of shares held by institutional investors increases. This almost seems implausible since institutional investors don't manage any of the R&D. However, what they do control is CEO compensation, and it appears that when institutional ownership increases, more of the CEO's compensation is tied to stock price. When that's true, CEOs themselves develop a longer-term perspective, increase R&D spending, and improve RQ.[7] Of course, few companies are lucky enough to attract institutional investors. What the other companies need is a way to fend off investor pressure to cut R&D to increase current profits.

Training Investors

In principle, companies should be able to make shareholders more patient merely by communicating how the increased investment in R&D will increase revenues, profits, and market value. Ultimately that will be true, but because RQ is new, shareholders need to develop confidence in these forecasts.

The best means to instill confidence is to present investors with short-run opportunities to profit from using RQ—trading strategies that allow sophisticated investors to benefit from the fact that only a few other investors know about RQ.

One simple but powerful strategy is merely to invest in an RQ50 Index fund. The RQ50 Index is the set of companies with the highest RQ in the prior fiscal year. While the RQ measure didn't exist until recently, the data to compute RQ go back to 1965, so I've constructed the index for every year from 1973 forward. I computed the RQs for all public companies conducting R&D in each year, ranked them in descending order of RQ, then identified the top 50 companies in each year. I then synthetically "invested" $20 in each RQ50 stock from fiscal year 1972 on January 1, 1973. On December 31, 1973, I "sold" all the stocks. I took the proceeds from that sale, divided that by 50, then invested that amount in each of the new RQ50 stocks on January 1, 1974. This is called an equal-weighted, annually rebalanced portfolio. Incidentally, about two-thirds of the stocks remained in the portfolio from one year to the next—reflecting the fact that R&D capability is fairly durable.

When I repeated this process annually through December 2015, I obtained the returns seen in Figure 8-2. What you can see is that the RQ50 Index fund outperformed the market by a factor of nine! In other words, if you had invested $1,000 in the market index on January 1, 1973, 43 years later (December 31, 2015) you would have $78,000. If, however, you had invested the same $1,000 in the RQ50 Index on January 1, 1973, you would have $708,000 on December 31, 2015![8]

FIGURE 8-2. **$1,000 invested in the RQ50 portfolio in 1973 would be worth $708,000 in December 2015 (versus $78,000 for the market portfolio)**

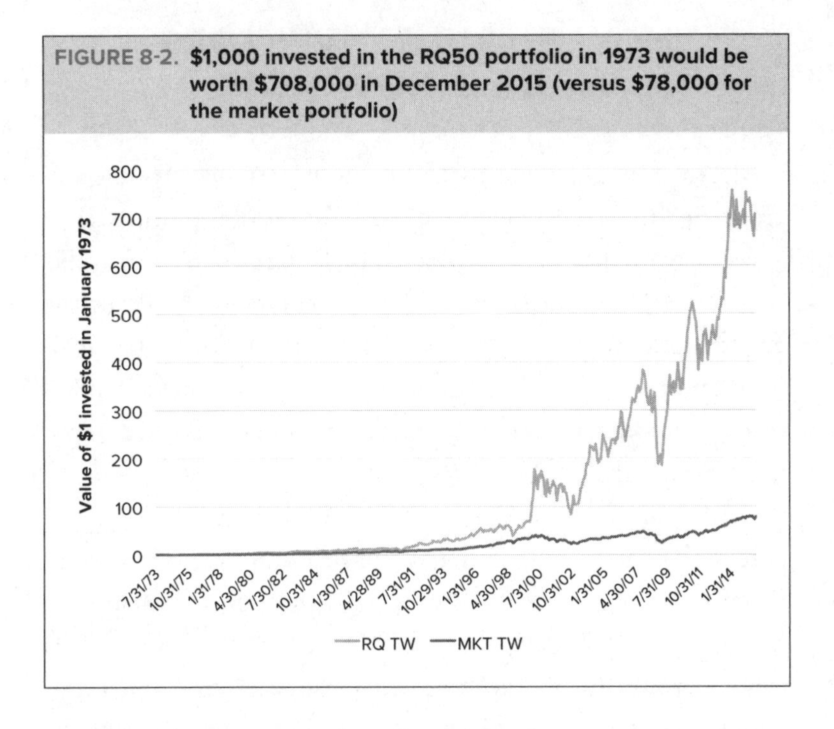

Why does the RQ50 Index perform so much better than the market? First, these stocks fly under investors' radar precisely because there hasn't been a reliable way to value innovation before. Ultimately (two to three years later) that R&D capability shows up in higher profits, which the market *does* know how to value. So the higher returns of the RQ50 Index are coming from investing in innovative companies before the market understands how their innovation will impact future profits.

The RQ50 Index is compelling evidence for both companies and investors that RQ delivers on its promise to translate R&D into market value. Next I'll show you what that translation looks like. Thus while the RQ50 should attract investor attention, the ability for investors to do fundamental analysis for all R&D stocks is what is ultimately needed for them to fully support companies' R&D investment. I discuss that next.

HOW *SHOULD* INVESTORS VALUE INNOVATION?

We learned in Chapter 1 that RQ determines how revenue and profit are affected by a company's R&D. A logical next step is to see how market value is affected by RQ. If financial markets are efficient, then the market value of a company should be the net present value of its future profit stream. Thus, if I forecast all future profits, discount them by the appropriate cost of capital each year, and add them together, I should obtain the company's market value.

This sounds easy in principle but can be complex to compute. Fortunately, we can make a few assumptions to clean up much of the complexity. The first assumption is that profits grow at a constant rate, *g*, forever. In fact, in the case of RQ, this isn't really an assumption—it's a result derived from a theory of growth discussed in Chapter 4. While the theory tells us that higher RQ translates into higher growth, it doesn't tell us by how much. We had to determine that from the data. A colleague and I found that a 10 percent increase in *R&D contribution* increases five-year growth 0.65 percent.[9]

The second assumption is that the cost of capital, *d*, remains the same forever. With those assumptions, we have a handy formula, the Gordon Growth Model (GGM), to make life simple. GGM says the present value (in this case market value), *V*, of a dividend stream (in this case profit stream) equals the annual income, π, divided by the cost of capital, *d*, minus the growth rate of profits, *g*:

$$V = \pi / (d - g)$$

This simple relationship explains where a stock's price to earnings ratio (P/E) comes from. It's merely $1/(d - g)$.

The next thing we need is profits. We know that profits equal revenues minus costs. In Chapter 10, I will show you the full profit equation for firms conducting R&D, but for now, I'm going to simplify that equation:

$$\pi = R\&D^{RQ} \times (\textit{Revenue contributions from non-R\&D inputs}) - \textit{costs}$$

Substituting the profit equation for π in the value equation yields the following equation for market value as a function of R&D and RQ:

$$MV = (R\&D^{RQ} \times (\textit{Revenue contributions from non-R\&D inputs}) - \textit{costs})) / (d-g)$$

What you see from this equation is that any time we increase RQ (keeping R&D constant), we increase the company's market value. The other thing you can see from the equation is that the contribution of R&D to revenues, profits, and market value combines two factors: the level of R&D and the productivity of that R&D (RQ) in a particular way: $R\&D^{RQ}$. I call this combination the *R&D contribution*. The *R&D contribution* amplifies the contribution from the other inputs.

You can of course refine this equation in additional ways. However, given the annual spread in minimum and maximum stock prices of even very stable companies, it's a little like using calipers to measure, and then cutting with a saw. The more interesting question is whether the market behaves as if RQ affects companies' stock price. The answer is yes. My colleagues Mike Cooper and Wenhao Yang and I compared companies' monthly stock market returns to their RQ, controlling for all other factors known to affect returns. We found that not only was RQ a significant predictor of returns, it was the *most significant* predictor. The next closest predictor was momentum.[10]

Trading Opportunity

Understanding the link between RQ and market value is the first step in making RQ interesting to investors—they can now value R&D in fundamental analysis. In fact, I thought as soon as RQ was available, it would rapidly diffuse throughout the investor community. I naively assumed that all investors preferred to do value investing, and therefore would jump on the opportunity to finally be able to value R&D.

It then occurred to me what investors must really care about is whether there is opportunity to trade on that knowledge. So far we've seen there is at least a trillion dollars in market value to be made if companies optimize their R&D budgets. However, only activist investors like Nelson Peltz are likely to get companies to optimize their R&D investment. An easier strategy, and the only strategy for most investors, is to identify companies that are undervalued given their RQ. This is precisely the strategy Warren Buffet uses for value investing in companies other than ones that do R&D. But how can you identify such companies? We saw one very simple way to profit from companies whose R&D was undervalued was to invest in the RQ50 Index.

When I saw that opportunity, I wanted to invest in the index myself! The problem was all my funds were tied up in a 401(k) at Vanguard. I approached our dean, showed him the returns to the RQ50 Index, and proposed that we create an "Olin RQ50 Index Fund." He was intrigued by the idea, so he put me in touch with Kim Walker, Washington University's chief investment officer. She too was intrigued, so she put me in touch with the head of product development at Vanguard. I described for him the foundations of RQ, and showed him the performance of the RQ50 Index. He first explained that the RQ50 is *not* an index. An index is intended to be a representative sample of companies in a particular domain. I asked, why would you want to invest in a representative sample when you could instead invest in exemplars? He then explained that even removing the name "index" from the fund wouldn't solve the problem, because Vanguard isn't interested in building funds around new theories. They let other investment companies experiment with new theories, and then pursue the ones left standing.

So the RQ50 Index is the simplest way to profit from investing in undervalued R&D companies. A second way that involves more work is to identify companies whose P/E ratio is below what it should be given their RQ. I showed earlier that P/E is merely shorthand for a company's expected growth and cost of capital. Since R&D investment is the primary driver of growth, then a company's price to earnings ratio (P/E) ought to track its *R&D contribution*. If the company's *R&D contribution* is rising (either because of increased R&D or increasing RQ), but the P/E is failing to track that, then the company may be undervalued (a buying opportunity). Conversely, if the company's *R&D contribution* is falling and the P/E is failing to track that, then the company may be overvalued (a selling opportunity).

In principle, the hardest step in determining if a company is undervalued using this approach is calculating its RQ. I've made that easy for investors. Access to a database of company RQs is available for free to all retail trading sites. Thus, RQ can appear on each stock page along with the standard metrics such as Market Capitalization, forward P/E, and Beta. If your preferred site doesn't provide RQ, recommend to it that it does.

Let's apply this approach to identifying opportunity using some case examples. Looking first at undervaluation, Merck's P/E tracked its *R&D contribution* pretty closely through 2008 (Figure 8-3a). Thereafter its *R&D contribution* increased,

FIGURE 8-3. **Using P/E ratio and R&D contribution to detect stock misvaluation**

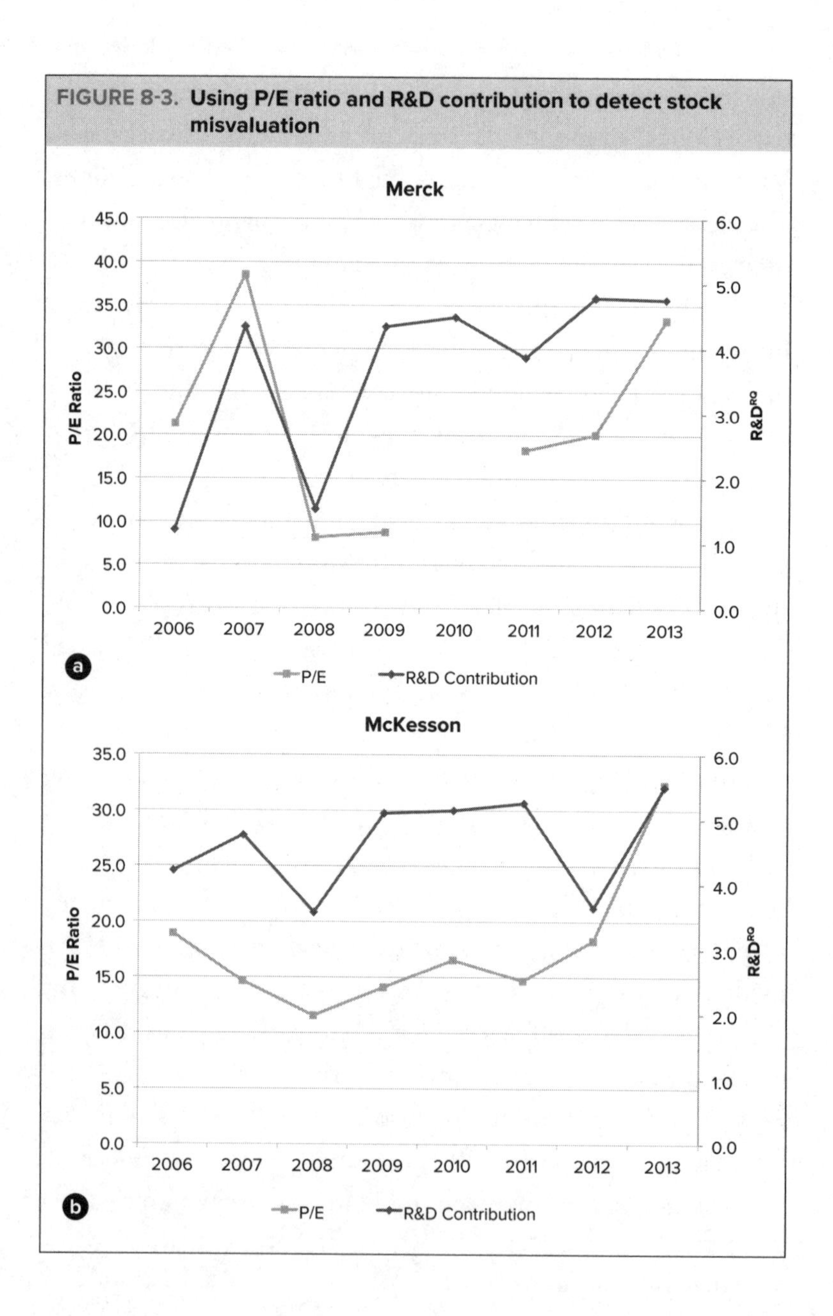

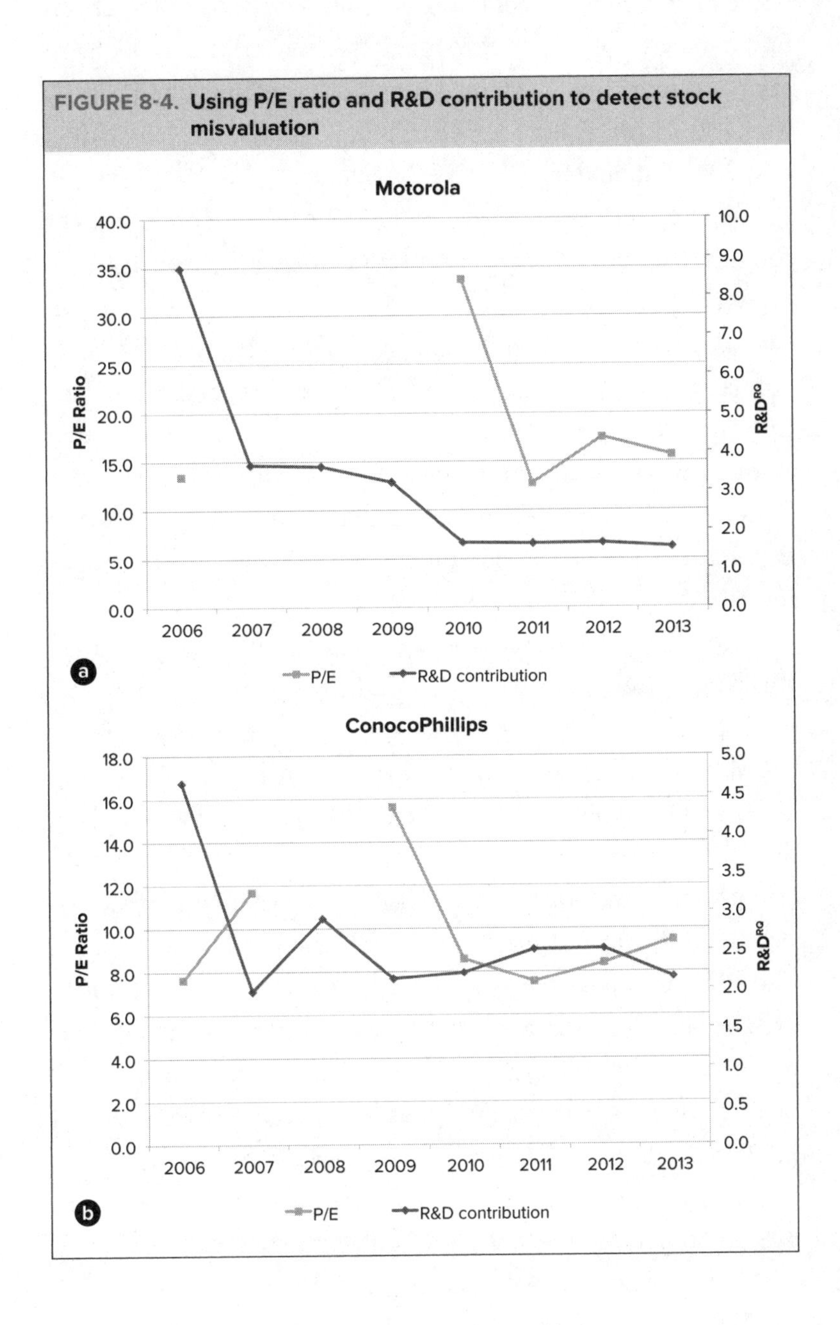

FIGURE 8-4. Using P/E ratio and R&D contribution to detect stock misvaluation

while its P/E failed to follow. Thus Merck appears to have been undervalued from 2008 through 2012. Indeed, the stock price seems to reflect that. Merck's stock price on January 1, 2009, was $26.75; its price four years later (January 1, 2013) was $44.20 (an annual return of 13.3 percent).

McKesson also appears to have been undervalued over roughly the same period (Figure 8-3b), and its returns were even more impressive over the same period. Its stock prices on the same dates were $44.20 and $105.23, respectively. Thus, its annual return was 24.2 percent. Note that several other factors affect these companies' stock prices over the period, such as market conditions and interest rates. This comparison of R&D contribution to P/E is merely one indicator a company is potentially undervalued, and therefore may warrant deeper investigation.

Merck and McKesson both appear to be undervalued via this comparison, but there is also trading opportunity for companies that are overvalued. This appears to have been the case for Motorola (Figure 8-4a) and ConocoPhillips (Figure 8-4b). In 2007 both companies' *R&D contributions* suffered a steep decline from the financial shock that never enjoyed the recovery of Merck and McKesson. While their R&D contributions remained low or fell further in 2008, their P/E didn't reflect that until two to three years later. Accordingly, both companies appear to have been overvalued—suggesting a signal to sell before the market recognized that. Indeed, Motorola's stock price on January 1, 2007, was $80.24, but fell to $17.91 on January 1, 2009—a 78 percent loss in value. Similarly, ConocoPhillips' stock price on January 1, 2007, was $50.63, but fell to $28.46 on January 1, 2009—a 44 percent loss in value.

A third way to exploit opportunities to trade on companies' RQ is to act on changes to either their R&D spending or their

RQ. This requires tracking a particular set of stocks, and checking them manually or setting up an alert system on a database that triggers your attention when either RQ or R&D is changing beyond some bounds.

Once you've identified such companies, the next step is to calculate the impact of those changes. To do that, you need to isolate the "non-R&D market value" (MV_{-R}) of the company. This is the market value of the company stripped of its R&D. To obtain that value, you need the company's market value in year t, before the change (MV_t), as well as its R&D and RQ before the change. The "non-R&D market value" (MV_{-R}) is merely the company's market value prior to the change (MV_t), divided by its R&D contribution prior to the change ($R\&D^{RQ}$):[11]

$$MV_{-R} = MV_t / (R\&D^{RQ})_t$$

The next step is to compute the new R&D contribution $(R\&D^{RQ})_{t+1}$ by changing either R&D, RQ, or both. Once you've done that, the expected market value after the change, (MV_{t+1}), is simply the "non-R&D market value" (MV_{-R}) times the new R&D contribution $(R\&D^{RQ})_{t+1}$:

$$E(MV)_{t+1} = MV_{-R} \times (R\&D^{RQ})_{t+1}$$

To make that a little clearer, let's apply this analysis to Motorola. In 2006, Motorola's market value (MV_{2006}) was $49.29 billion. Its R&D was $3.68 billion, and its "raw RQ" was 0.264. So Motorola's R&D contribution was 8.74 (= $3{,}680^{0.264}$). We obtain its "non-R&D market value" $(MV_{-R})_t$ by dividing its market value by its R&D contribution. This value is $5.64 billion (= $49.29 billion/8.74).

In 2007, Motorola underwent two changes. It increased its R&D to $4.14 billion, but its "raw RQ" decreased to 0.156, so its new R&D contribution was 3.67 (= $4{,}140^{0.156}$). When we multiply Motorola's new R&D contribution (3.67) by the "non-R&D market value" (MV_{-R}) computed using 2006 data ($5.64 billion), we obtain Motorola's expected market value for 2007 of $20.7 billion (a 58 percent reduction):

$$E(MV)_{2007} = MV_{-R} \times (R\&D^{RQ})_{2007} =$$
$$\$5.64\ billion \times 3.67 = \$20.7\ billion$$

Thus knowing Motorola's R&D and RQ would have allowed investors to anticipate a decrease in the company's stock price, even though the company was increasing its R&D.

A final strategy to trade on a company's RQ is to invest in anticipation of its RQ changing. This requires both knowing that a company's R&D strategy has changed and recognizing how that strategy change will affect its RQ. We are still pretty far from being able to do that. Fortunately, for the time being, there is plenty of opportunity to pursue the undervalued/overvalued investment strategy, and the strategy of responding to observed changes in R&D or RQ.

Relating This Back to Companies

We've gone pretty far down the path of helping investors trade on opportunities that exploit the fact that most other investors don't know how to value R&D. You can now exploit those opportunities for your personal investing. That wasn't our primary goal for this chapter, however. Our primary goal was to alleviate the current problem companies face from investors—unwarranted pressure to reduce R&D.

The role of the trading strategies was to demonstrate that investors ultimately should value R&D correctly, because those that do can make money in the short run by trading against those who don't. Ultimately, this will drive the don'ts from the market. When that happens, companies should be able to pursue R&D strategies they're currently dissuaded from. In the meantime, you may want to court institutional investors, since they already seem to value R&D.

SUMMARY

The first way RQ helps companies is by telling them how much they should invest in R&D. However, if investors don't know how the increased investment translates to market value, companies may not have the discretion to expend additional funds. Fortunately RQ defines the relationship between companies' R&D and their market value. While I showed that relationship, I also showed that because most investors don't know companies' RQs, there is trading opportunity for those who do! Ultimately, of course, all investors should know companies' RQs. So the bad news is the trading opportunity will go away, but the good news is once that happens, companies will no longer be under pressure to cut R&D expenditure (unless of course their R&D exceeds the optimum).

Given the influence investors exert over companies, they too play a critical role in restoring economic growth. Even though they don't invest in R&D directly, they influence the behavior of companies that do. The proof that investors affect behavior is that R&D spending and RQ increase when companies have more sophisticated investors (institutional investors).

THE PROMISE OF RQ: Restoring Growth

In Chapter 4 we learned that RQ is immediately useful to companies in determining more optimal R&D budgets. There is also a greater, though longer-term, promise of RQ—that companies can improve their R&D capability and restore their earlier growth. Just as TQM helped companies improve manufacturing efficiency, hospital report cards helped hospitals improve morbidity and mortality, and sabermetrics helped teams improve recruiting and ultimately player skills, the promise of RQ is that it can help companies improve their innovation capability.

This of course is only possible if RQ can change. The key question then is "Can it?" We know from Chapter 1 that RQ can change in the *wrong* direction (the 65 percent decline over the past few decades), but is it possible for companies to change in the right direction—increase their RQ? The good news is the answer appears to be yes.

Figure 9-1 compares companies' RQs over two 10-year periods. Each dot represents a company. Drawing a vertical line

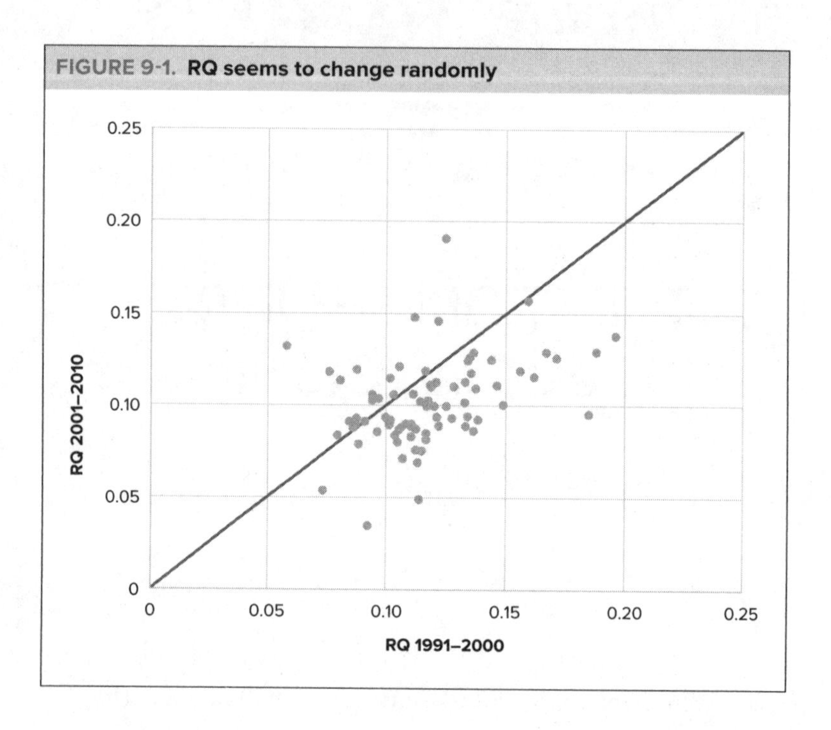

FIGURE 9-1. **RQ seems to change randomly**

from any dot to the x-axis tells you that company's RQ from 1991–2000; drawing a horizontal line to the y-axis tells you that company's RQ from 2001–2010. The diagonal line represents the constant RQ line—which is where all companies would fall if their RQ didn't change. This is clearly not the case. The figure indicates that RQ changes substantially, but there is no systematic direction to that change. This is precisely what we would expect from companies engaged in trial and error learning without feedback.

While there is no systematic direction to the change, the alarming news is that only 25 percent of companies improved their RQ (those above the diagonal), while roughly 75 percent of companies saw their RQ deteriorate (those below the diagonal). Thus RQ can change either through deliberate efforts to improve or, more likely, through inadvertent factors that lead to decay. Thus it is highly likely a company's RQ *will* change. The challenge then is to drive positive change and avoid negative change. This is precisely what we hope to accomplish in this chapter.

WHY HAS RQ DECLINED?

The first step in understanding the potential to improve RQ is to understand why RQ has declined so dramatically. This is important for two reasons. First, it may be the case that there are external forces making R&D more difficult. If so, then the opportunities to improve RQ may be limited. If this isn't the case, then it is likely the decline is self-inflicted. Thus, the second reason to investigate the decline is to understand what companies did to induce it. If we understand that, we may be able to improve RQ merely by reversing those company behaviors.

One possible explanation for the fact that RQ declined for the majority of companies is that R&D has gotten harder. This is the theory proposed by Chad Jones[1] that we discussed in Chapter 4, when we were trying to understand why increasing R&D doesn't increase innovation. You may remember that Jones identified two mechanisms through which this would occur. The first mechanism is a "fishing out" (or cherry-picking) effect: the notion that the most obvious ideas are discovered first, so that the quality of remaining ideas is degrading over time. The second mechanism is diminishing returns to research labor: the idea that adding more researchers decreases the number of innovations per worker because it increases the likelihood researchers are duplicating one another's efforts.

Another theory consistent with the idea that R&D has gotten harder comes from Ben Jones in "The Burden of Knowledge and the Death of the Renaissance Man."[2] Ben's idea is that as knowledge accumulates, our ability to move the frontier requires new generations to first learn what is known from prior generations. Perhaps the best way to think about Jones's theory is to consider knowledge as an ever-expanding circle, and points along the perimeter as the frontier in a given domain. If a researcher requires a given volume of knowledge to innovate, and if the researcher must start at the center of the circle, then as the knowledge circle grows, new researchers face a strategic choice. They can either narrow their area of expertise (wedge of the circle), or take more time to get to the frontier for a wider wedge of the circle. Indeed, Ben found that the age of scientists at the time of their first invention, the average level of research specialization, and the number of coauthored innovations were all increasing over time. This indicates that getting to the frontier takes longer and, even at that, requires researchers to specialize in a narrower range of knowledge. Further, because these ranges of knowledge are smaller, researchers need more coauthors to cover enough knowledge to create a new discovery.

What's dismal about both the Joneses' "R&D getting harder" theories is that if they're correct, there is little hope that companies can improve their RQ. However, you may remember from Chapter 4 that I presented results from a test of whether that was true. The idea behind the test was that if R&D has gotten harder, then it should be harder for all companies. Accordingly, not only will *average* RQ decline (as we saw in Chapter 1), but *maximum* RQ should decline as well. However, when I examined over 40 years of RQ data, I found that maximum RQ was *increasing* across the economy and sectors. Thus there was little support for the idea that R&D has gotten harder. However, when I dug deeper, I found that RQ was *decreasing* within industries (shown previously in Figure 4-1).

The implications of the pattern are actually pretty exciting. What the pattern suggests is that while opportunities within industries decline over time, as they do, companies respond by creating new industries. Once I saw this pattern it was easy to think of examples. In fact, many of these examples are referenced in the current debate on disruption. Some common examples are the death of the typewriter and its replacement by personal computers, and the death of landlines and their replacement by cell phones. While there are numerous other examples, what is true in the two cases that came to mind is that the market for the new technology was actually much broader than that for the technology it replaced. As an example, personal computers enjoy an installed base in the United States of 310 million machines,[3] while the installed base of electric typewriters was only 10 million machines at its peak in 1978.[4] While we can't say whether this pattern of increasing maximum RQ will continue forever, it has persisted over the 40 years for which we have data.

The good news is that while industries may be doomed, companies don't have to be. They can move into industries with greater opportunity, while exiting industries with declining opportunity. That news also provides our first lesson in improving RQ. Companies likely have to diversify to avoid diminishing opportunities in their own industry. This general pattern is at least a century old, and in fact was the genesis for industrial R&D. This genesis is captured in vivid case histories of DuPont, General Motors, and Standard Oil in Alfred Chandler's book *Strategy and Structure*.[5]

This prescription to diversify to avoid declining opportunity is similar to the broad prescriptions in each of the prior chapters. Over the set of all U.S. companies, these tend to be strategies that improve companies' RQ. However, not all strategies work for all companies, so what you want to understand is what works for *your* company. Knowing what works for your company requires the ability to diagnose problems within your company.

Diagnosing Decline Within Your Company

The key to diagnosis is data. This is the principle behind the burgeoning data analytics movement. The Internet has made enormous amounts of data available to companies. Some of these data are used to improve operations of existing companies—think of Amazon tracking your purchasing habits to recommend other purchases. In other cases, new companies have been created to make these data available to other organizations. PASSUR Aerospace, for example, uses "big data" to predict aircraft arrival times. This ability is so important to commercial air carriers that PASSUR's analytics are used to manage 53 percent of all domestic flights in the United States.

There are two principles behind the use of data to enhance performance. The first is the simple principle that conspicuousness motivates initiative. I'm driven to perform better when I know others can view that performance: "What is measured is improved."

This is a nontrivial component of the improvement in restaurant hygiene that Ginger Jin and Phillip Leslie observed when Los Angeles County implemented a system in 1998 for grading the hygiene of all restaurants.[6] Ginger and Phil were interested in understanding first, how responsive consumers were to the new information and second, how consumer response translated into improved restaurant quality. They found not only that customers began frequenting restaurants with better scores, but that restaurants responded by improving their hygiene levels. Moreover, there was an unanticipated benefit: fewer instances of food-related illnesses at Los Angeles hospitals.

Other examples of conspicuousness leading to improvement are the No Child Left Behind Act, which resulted in higher mathematics achievement, particularly for students from disadvantaged backgrounds, and New York's Cardiac Surgery Reporting System (CSRS), which reports which hospitals performed better or worse than the statewide average. David Cutler, Rob Huckman, and Mary Beth Landrum found that these hospital "report cards" led to decreased mortality, particularly among low-performing hospitals. However, they also found that the dark side of report cards was that hospitals began rejecting patients with higher mortality risk.[7]

I refer to this mechanism of merely reporting scores as "conspicuousness" because in all these cases the bulk of the improvement came from low-performing organizations. This implies no new knowledge or analysis was necessary to achieve the gains. Rather, substantial improvement could occur merely by adopting practices already in effect at higher-performing organizations.

A compelling study examining what drives improvement in the presence of these ranking systems was conducted in the context of hospital emergency departments (EDs).[8] Hummy Song and Anita Tucker compared two systems for ranking physicians on patients' average length of stay (time from patient login to discharge) in the ED. In the private system, physicians only learned their own rank relative to all other physicians. In the public system, physicians learned not only their own rank, but also the identities and rank of all other ED physicians. Hummy and Anita found the private ranking system had no impact on patients' lengths of stays, but that the public ranking system reduced length of stay an average of 17 minutes. By digging deeper, they found that neither motivation to be top-ranked nor shame of being bottom-ranked seemed to drive improvement. Instead, the mechanism driving the improvement was the ability to identify the exemplar physicians and ask them about their practices. Thus the first thing companies can do to improve their RQ is to identify who among their rivals has the highest RQ.

The other mechanism for data-driven improvement is more sophisticated, and it predates big data analytics by a century. This form of improvement utilizes data for three purposes: first to baseline performance, second to diagnose its root causes, and third to conduct experiments and/or analysis to identify solutions.

The birth of this movement is captured in the 1911 book *The Principles of Scientific Management* by Frederick Taylor.[9] Taylor's goal for the book was to reverse what he saw as mass labor inefficiency that caused the entire country to suffer. He argued that the source of the inefficiency was use of rules of thumb in executing tasks. He said that even though the "best practices" in each of the trades had evolved over many generations, where presumably each generation improved over the prior generation, there was no uniformity in the way tasks were executed from one geographic region to another. While uniformity was not the goal in and of itself, lack of uniformity demonstrated that the centuries-long evolution did not converge on an optimal practice. Otherwise all regions would have had the same practice.

Taylor advocated replacing rules of thumb with "scientific management" in which the knowledge embedded in the tradesman's practices was codified, analyzed, and translated into "rules, laws, and formulae." His most compelling example of the technique examined pig iron handlers—steelworkers whose job was to pick up 92-pound "pigs" of iron that had been stored in piles and load them onto railcars. At the start of his investigation, an average pig iron handler was processing 12.5 tons per day. After running experiments and analyzing the results, Taylor discovered that the factor limiting how much pig iron a handler could process was not "work" (in the physics sense of mass times distance); rather, it was the time the pig handler was bearing the weight. This realization led to job redesign in which, among other elements, the optimal periods of load-bearing versus resting were identified. The changes in task design led to *quadrupling* the amount of pig iron a handler loaded in a day. Another impressive example was bricklaying, a craft that had evolved over centuries. Prior to scientific management, the best bricklayers could lay 120 bricks per day; after task redesign, they could lay 350 bricks per day. Similar gains were shown for other trades.

At the time Taylor's book was published, he claimed that 50,000 U.S. workers (of approximately 8 million manufacturing employees) were governed by scientific management systems. In those companies utilizing the system, average output had doubled and wages had increased 30 percent to 100 percent. Thus, both companies and employees were better off.

"SCIENTIFIC MANAGEMENT" APPLIED TO R&D

As with scientific management, the first step to improving RQ is simply knowing your RQ. You will learn how to do that in the next chapter. Once you know your RQ, the second step is to benchmark it. There are three forms of benchmarking you want to conduct: (1) compare your RQ to rivals, (2) compare your current RQ to your prior RQ history, and (3) compare RQs across your company's divisions. Each form of benchmarking has distinct implications for improving RQ, so we discuss them in turn.

RQ Versus Rivals

The first form of benchmarking is to compare yourself against rival companies. The role of this comparison is to determine whether there is "better practice" within your industry. To the extent there is a company with a higher RQ, there is opportunity to improve your company's RQ merely by adopting some of that company's better practices. This is a high payoff form of benchmarking because the answer to the question whether there is better practice in your industry is yes for all but one company. This form of benchmarking tells you three important facts, (1) that there *is* room to improve, (2) *how much* room there is, and (3) *who* you should look to for guidance on how to improve. Interestingly, merely knowing where you stand doesn't lead to improvement. As we learned for the emergency department study, you also need to know the identity of the high RQ company. This allows you to know whose practices to emulate and ideally gives you an opportunity to approach them about what they do differently.

In addition to looking at your rivals' current RQs, you can also track their RQs over time. One reason to do so is to determine if your industry is one that is now in decline. Since we now know that opportunities within industries decay and new industries emerge to replace them, observing an industry trend where all companies' RQs are declining may be a signal it's time to exit the industry. Note however that observing decline is not necessarily a sign opportunity is diminishing; it may just mean all companies in the industry are merely "following the leader" in adopting inferior R&D practices.

Current RQ Versus Prior RQ

The second form of benchmarking is to compare your current performance to your past performance. Since we learned in Chapter 1 that, on average, companies' RQs have declined 65 percent, this too is a high-payoff form of benchmarking. By tracking your own RQ over time, and particularly looking for dramatic changes, you may be able to tie changes in your RQ to changes in your R&D practices. If you can identify practices that have changed around the time RQ began to decline, you have candidates for improving current practice.

One example from the cases we've seen so far is the decline in P&G's RQ from 2000 to 2007. Looking back through changes that occurred around 2000 suggests that decentralization of R&D was a feasible explanation for the decline, and accordingly might be a candidate for reversal.

Another example from the prior cases is the decline in HP's RQ following the departure of Dave Packard as CEO. One thing that changed was substantial acquisition and divestiture activity. HP sold off its core, Agilent, in November 1999, and replaced it with acquisitions of existing companies, most notably Compaq in 2001 and EDS in 2007. But the acquisitions alone don't account for the new RQ. Quotes from Charles House in the Nocera article suggest there was also disinvestment in tools and equipment (which fits with the decline in R&D intensity).

RQ Differences Across Business Units

Perhaps the most valuable form of benchmarking is to compare RQ across your company's business units or product lines. The advantage of intracompany differences as a means for diagnosis is that once you identify where best practice lies within the company, you can easily access it. When best practice lies outside the company, you may have no way of learning what the rival does differently.

We found that business unit RQs differ substantially, with the RQ of the highest business unit being as much as 30 points above the lowest business unit. This is equivalent to being in the top sixth of companies versus being in the bottom sixth of companies.

Once you have identified where best practice (highest RQ) lies, either with a rival, in your own past history, or in a particular business unit in the company, the third step in applying scientific management to innovation is identifying what practices underpin that RQ. If best practice lies within the company (either past history or within one of the business units) this is relatively straightforward, though it may be cumbersome. What you want to do is catalog all the R&D practices in place currently across business units or rivals, and in the case of a higher prior RQ, you also want to characterize how those practices differed at various points in time (as I did preliminarily for P&G and HP). The goal is to pinpoint which differences in practices are driving the RQ differences.

In those cases where the higher RQ lies outside the company, identifying the factors driving the higher RQ becomes more difficult. A starting point is to gather as much publicly available data as you can on how that company conducts R&D. This may involve examining the patent and publishing records. Alternatively, it may be as simple as attending conferences and merely asking colleagues at rival companies about what they are doing.

Once you have a set of candidate practices that appear to differ between your current practice and other, higher RQ practices, you want to analyze them to understand the cost to implement them and the likely impact of implementing them.

In the absence of large-scale quantitative evidence like that derived from my NSF study and documented in this book, each of these candidate practices is merely a hypothesis about what drives RQ. Accordingly, you need to implement the practices in gradual fashion. In essence you want to conduct trial and error experiments to test their efficacy. Fortunately, you now have a tool, RQ, to provide feedback on the efficacy of those experiments.

THE PAYOFF OF IMPROVING RQ

In the previous chapter discussing Wall Street, I said that companies could improve their market value first by rightsizing their R&D, and second by improving their RQ. We also learned how both strategies increased companies' market value. The payoff of rightsizing R&D for the 63 percent of companies that are overinvesting in R&D is an immediate gain in profits by eliminating the overinvestment. The payoff for the 33 percent of companies that are underinvesting is a slightly delayed gain in profits from increasing R&D 10 percent. These payoffs of rightsizing R&D are substantial, on average $182 million dollars in forgone profits per company. However this payoff is small relative to the payoff of increasing RQ.

Cases

Now that you understand how to diagnose your innovative capability when you *do* have a measure of R&D productivity, let's step back and see what happens when you try to diagnose your innovative capability when you *don't* have such a measure. I'm going to present case studies of two companies, each of which undertook bold changes in its R&D strategy: Trimble Navigation and AT&T. As you read each case, try to determine whether the strategy improved or degraded the company's RQ.

Trimble Navigation

Trimble Navigation was founded in 1978 to produce positioning and navigation products. Through the 1990s Trimble developed and patented a number of technologies, reaching a single-year peak of 94 patents in 1997. In addition, the "Company History" on Trimble's website indicates it was rapidly expanding the set of product markets in which this technology was deployed. In 2000, however, it appears the company changed its strategy from one of internal development to acquisition. Up until 2000 each paragraph described a technological development; from 2000 forward there is no mention of developments, only of

FIGURE 9-2. **Trimble Navigation patent and acquisition trends**

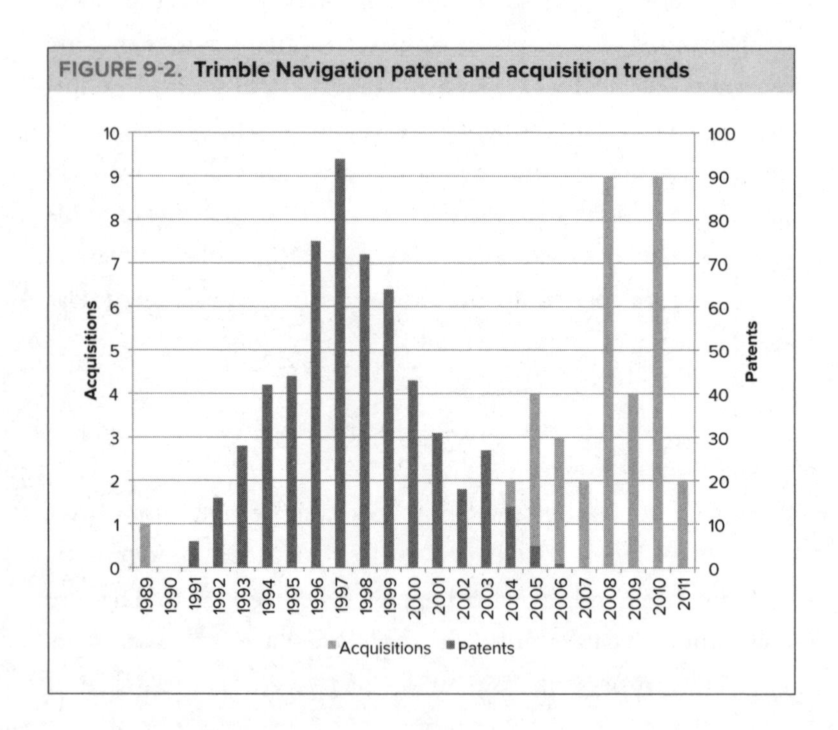

acquisitions.[10] This strategic shift is also captured by comparing Trimble's patent rate to its acquisition rate (Figure 9-2). Acquisitions are rising, while patents are dropping. A shrewd analyst might pick up on either or both of these trends. However, without the RQ measure it would be difficult to tell whether the shift was value-enhancing or value-destroying.

AT&T

AT&T is over 120 years old as well as one of the largest companies in the world. Its roots date back to Bell Telephone—the company founded by Alexander Graham Bell to commercialize his innovation. The company currently known as AT&T was the former Southwestern Bell (SBC), one of the regional Bell operating companies (BOCs) created in the court-ordered divestiture of American Telephone and Telegraph in 1983. SBC acquired its former parent for $16 billion in 2005, and took on its branding.

AT&T has three operating segments: Wireless (49.8 percent of revenues), Wireline, which includes traditional voice, U-verse TV/broadband and commercial networks (47.2 percent of revenues), and Advertising Solutions (Yellow and White pages). While much of the company's growth can be attributed to wireless and data penetration in the market generally, AT&T has also engaged in innovation. Innovation on the customer side is geared toward increasing data usage for existing customers as well as inducing new customers to switch from other carriers. Innovation on the network side is geared toward cost efficiency in accommodating the demand growth.

At the heart of AT&T innovation is AT&T Labs. The seed for the Labs was a set of researchers who were left behind when Bell Labs went with Lucent as part of its spin-out from AT&T in 1996. While the labs were originally unconstrained in the direction of innovation, they are now more closely tied to AT&T's segments: commercial and networking technologies, products, and services.

Innovations from the Labs deemed to have commercial potential are moved to "AT&T Foundry" in one of three locations (Israel, Dallas, or Silicon Valley). AT&T claims the foundry commercializes new products and services three times more quickly than normal.

In addition to the primary innovation flow from the Labs, AT&T has two other sources of innovation: "Fast Pitch," an effort to incubate entrepreneurs that want to collaborate with AT&T (approximately 500 pitches per year), and "The Innovation Pipeline," an internal idea sourcing website open to all employees (14,000 ideas submitted to date).

This innovation flow appears to have escalated over the past decade. Exhibit 9-3 shows the patent rate has grown from 40

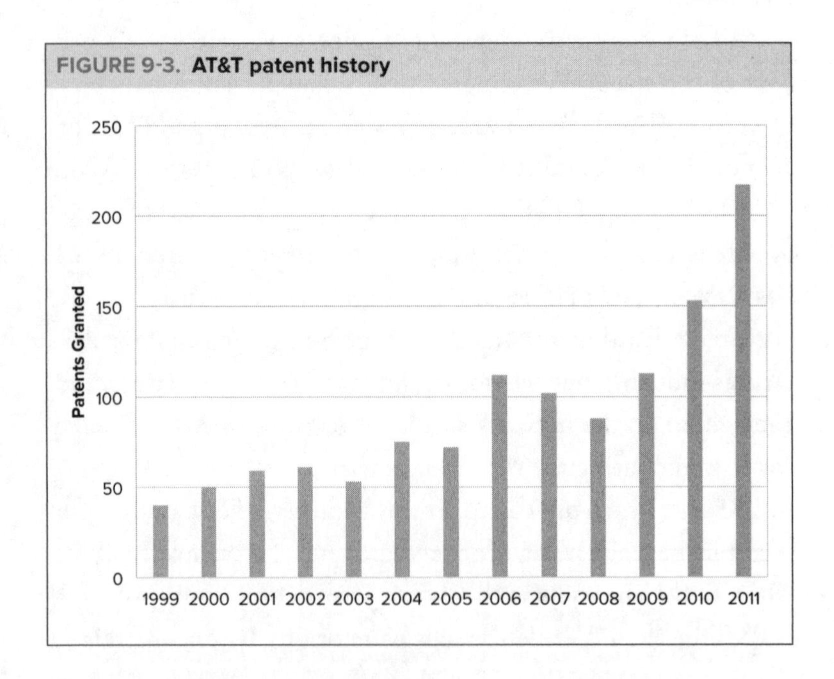

FIGURE 9-3. AT&T patent history

patents in 1999 to 217 in 2011. Moreover, AT&T's innovations appear to come primarily from organic processes. While there have been acquisitions, these have been principally to gain spectrum—the most binding constraint on AT&T's growth.

Verdicts

You've seen both Trimble Navigation and AT&T. Now take a stab at whether each company's strategy exhibited increasing or decreasing RQ over the period we discussed. If you guessed that AT&T's strategy *increased* its RQ, give yourself one point. If you guessed that Trimble's strategy *decreased* its RQ, give yourself a point. If you guessed both correctly, give yourself a bonus of 3 points. We now present each company's RQ and financial performance in an effort to show how each strategy affected both.

Trimble

Figure 9-4 indicates Trimble's RQ was increasing through 2004. In 2004, however, Trimble's RQ fell 40 percent. It's net income tracked that with a lag, growing fairly rapidly from 2000 to 2007, but then suffering a steep decline in 2009. Thus Trimble's RQ signaled the company's ascent and decline four years before they materialized in net income. Admittedly, a substantial portion of this profit decline was due to the recession. Trimble's decline of 54 percent is on par with the 62 percent average decline for U.S. publicly traded companies. What differs for Trimble, however, is that while the average net income for U.S. companies is now 8.5 percent higher than the prerecession peak, Trimble's net income remains 27 percent below its peak. Thus something beyond the recession is contributing to the decline.

FIGURE 9-4. **Trimble RQ history**

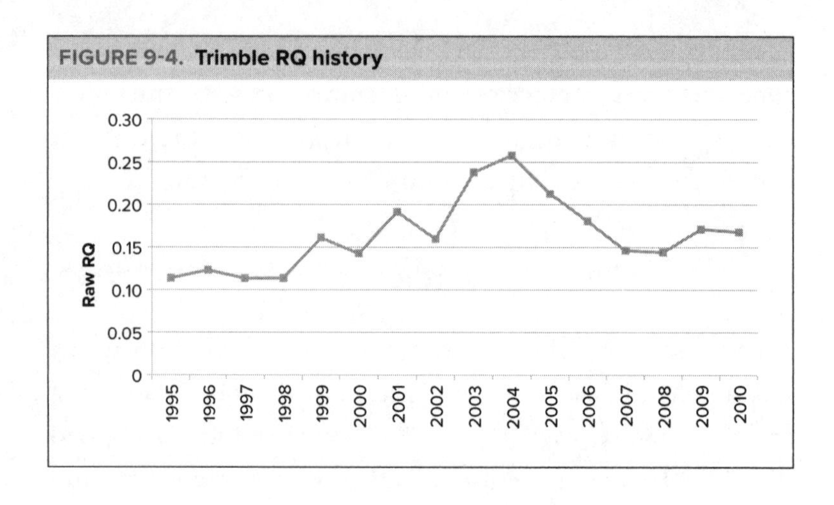

An additional tack on whether RQ is capturing real change examines Trimble's R&D spending—does the patent trend mean it abandoned R&D? The answer is no. Trimble continued to grow R&D spending, but not sufficiently for the acquisitive growth. Knowing Trimble's RQ allows us to compute its optimal R&D spending and to compare the optimum to what Trimble actually spent. That comparison (Figure 9-5) indicates Trimble actually overinvested on R&D through 2002. Thereafter it has been underinvesting even for its lower RQ.

Note the fact that Trimble had no patents in the later years in and of itself does not imply low RQ. We know companies can have high RQ without patenting. However the fact that Trimble had a prior record of heavily patenting suggests that patents are valuable in its industry. Accordingly the lack of patents likely reflects a change in innovation strategy rather than a change in patenting strategy.

This cursory analysis is based solely on publicly available information. If this were an internal analysis, there would likely be several more candidates for the decline in RQ. Nevertheless,

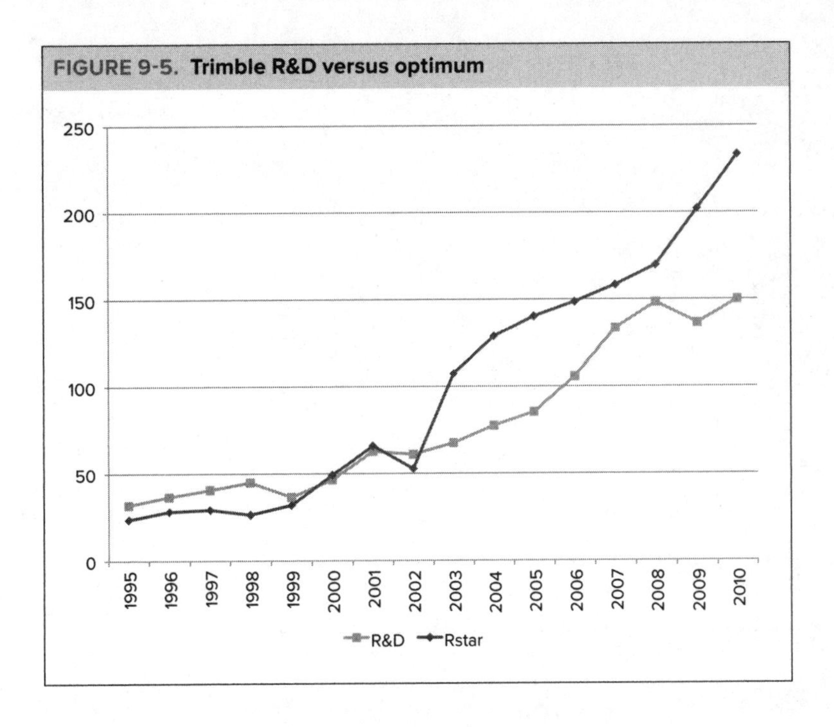

FIGURE 9-5. **Trimble R&D versus optimum**

the case of Trimble Navigation illustrates the predictive power of RQ. It also illustrates how maintaining a high RQ and managing R&D investment commensurately has significant impact on bottom line results. Finally, it indicates that one way companies lose RQ is by substituting external R&D for internal R&D.

AT&T

I chose AT&T as the second case for changing RQ both because it has an impressive 16.6-point increase in RQ (Figure 9-6) and because its net income grew from $7.3 billion in 2006 to $19.9 billion in 2010. While this net income growth includes acquired growth of $4.5 billion from the addition of Southern Bell in 2007, this still represents 66 percent growth over the remaining three years.

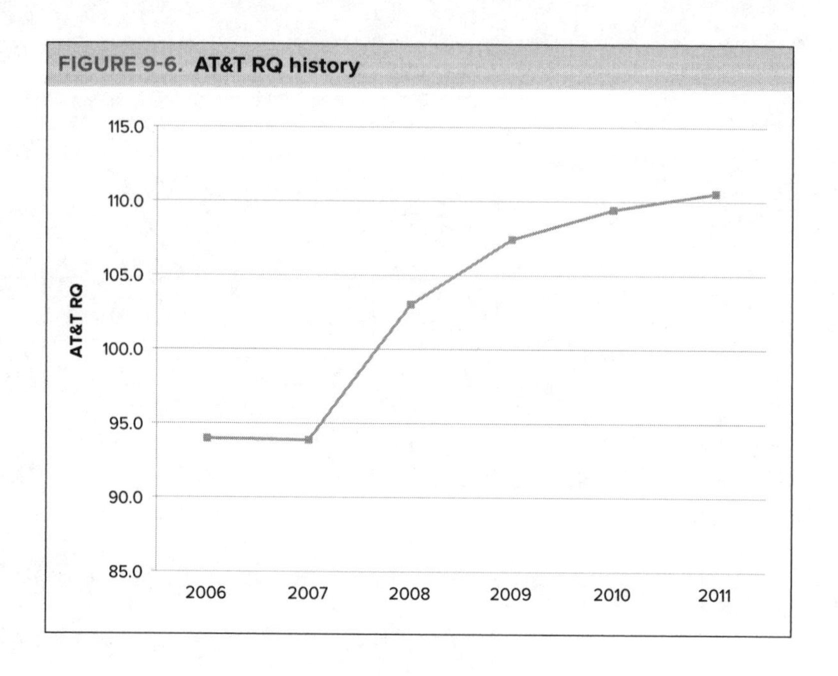

FIGURE 9-6. AT&T RQ history

Avoiding Trimble's Fate

Both the AT&T strategy and the Trimble strategy seem plausible up front. It's easy to see how the strategies got backing both internally as well as from investors. You may have other R&D strategies within your own company that seem equally plausible. How do you know if you're on the AT&T path or the Trimble path? You don't, *initially*. The problem then is not choosing bad strategies—the whole point of R&D is experimentation. The danger is failing to pull the plug on a bad strategy. In the past there wasn't a good way to do that.

Everyone has trouble pulling the plug. We saw earlier this problem kept VCs from earning three times the returns they currently enjoy. The biggest problem in both the VC case and the R&D case is *knowing when* to pull the plug. Both environments are plagued with tremendous uncertainty. RQ reduces that uncertainty. It provides a tool for knowing when to pull the plug.

SUMMARY

The Trimble and AT&T cases indicate that RQ *can* change—in the case of Trimble (as in 75 percent of companies), it changed for the worse; in the case of AT&T, it changed for the better. Both companies implemented plausible R&D strategies. However, because the companies lacked feedback, they had no way of knowing whether the strategy was effective. Now that we have RQ, companies can tell whether their strategy is indeed sound.

Trimble and AT&T are isolated examples, but they demonstrate how substantially companies can change their innovative capability, and they offer some clues for how to change it in a favorable direction. Very few companies look exactly like Trimble or AT&T, however. Thus we neither want to mimic AT&T nor avoid mimicking Trimble. Either strategy might be a mismatch for your company. Fortunately, the other chapters have identified strategies that have been tested over the full spectrum of companies. Those strategies offer more reliable prescriptions for what companies should change to improve their RQ.

To determine more specifically what R&D practices to adopt in your company, we recommend a "scientific management of R&D" approach involving three steps: (1) learn your RQ, (2) benchmark your RQ against rivals, against your own history, and across divisions to identify who represents best practice (highest RQ), and then (3) investigate what that best practice consists of.

What's rewarding for companies and their shareholders about improving RQ is that it will increase revenues, profits, and market value for any given level of R&D. More exciting, however, is that if all companies restore their RQ to prior levels, we should be able to restore economic growth to the levels Robert Gordon hails in *The Rise and Fall of American Growth*.

BEHIND RQ: What It Really Is and How to Find Yours

We've seen many ways RQ can help companies generate greater value from their R&D. Until now, however, you have had to take it on faith that RQ is a meaningful and valid measure. This chapter provides the evidence to replace the faith. It describes the RQ measure, discusses how it is derived, links it to economic theory, and shows through that theory how RQ can predict optimal R&D spending, revenue, profits, and market value.

THE THEORY AND METHOD BEHIND RQ

I argue that RQ is the most intuitive measure you could construct for R&D effectiveness (though once you see the equation for RQ, you may not agree). It's so intuitive it occurred to me as a first-year PhD student. In fact, my being a novice may be the reason RQ occurred to me rather than someone else—I hadn't yet become steeped in the other measures. The problem at the time was I couldn't act on the intuition because there weren't yet statistical tools that allowed me to estimate RQ for companies. Even when those tools first became available 10 years ago, it took the computer program all night to run the estimation.

The reason RQ is so powerful is that it is rooted in economics. It's derived from the standard production function, which defines the relationship between company inputs and outputs, including the trade-off between inputs. The production function can be expressed as Equation 1 (*on the next page*), or shown graphically, as in Figure 10-1. The horizontal axis in Figure 10-1 is the amount of input B (let's call it labor), while the vertical axis is the amount of input A (let's call that capital). Each curve represents a particular level of output. Let's take the innermost curve, and let's assume it represents 100 widgets of output, and that the other curves represent 200 widgets and 300 widgets. Each point along the inner curve corresponds to a combination of A and B that will generate 100 widgets, so the curve is called an *isoquant*. You can see, for example, that you can produce 100 widgets by using a lot of A and very little B (point b), or about equal amounts of each (point d). The question you might ask is how to decide where on the isoquant to be (what combination of A and B is most cost effective). To answer that question, you need to have information about the prices of A and B. That's what the diagonal line represents—the relative prices of A and B. In this example, A and B have the same price. The most cost effective combination of A and B for any given level of output is where the price line intersects the isoquant for that output. In the 200-widget example, the price line and isoquant intersect at point c, so we would use equal amounts of A and B.

FIGURE 10-1. Isoquants show a production function graphically

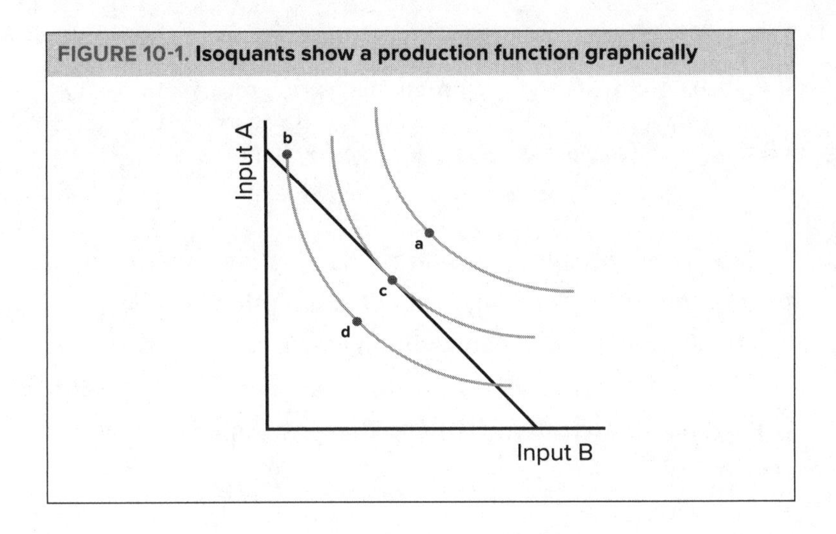

We are going to come back to this intuition when we talk about the optimal levels of R&D, but for now I want to focus on the equation for the production function. Equation 1 captures the shape of the curves in Figure 10-1. What the equation tells us is that the number of widgets we produce (output) is determined by the amount of each input (capital, labor) and by the productivity or "elasticity" (α, β) of each input. Each elasticity has a meaningful interpretation. It's the percentage increase in output associated with a 1 percent increase in that input (holding everything else constant). As an example, if I increase labor 1 percent, and keep everything else the same, output should increase β percent.

(1) $$Output = Capital^{\alpha} \times Labor^{\beta}$$

The production function used to generate RQ differs from Equation 1 in two ways. First, it expands the equation to include intangible inputs: *R&D, spillovers, and advertising* (Equation 2). Equation 2 itself is not new—the basic "R&D production function" has been used by economists for decades.

In fact, the elasticity on R&D, γ, is the most common means used by economists to measure industrywide returns to R&D.[1]

(2)

$$Output = Capital^{\alpha} \times Labor^{\beta} \times R\&D^{\gamma} \times Spillovers^{\delta} \times Advertising^{\phi}$$

Thus the only innovation to the production function used to generate RQ is changing one key assumption. While economists assume all companies in an industry share the same elasticities, I assume they differ across companies. So each company (i) has its own elasticity for each input (Equation 3):

(3)

$$Output = Capital^{\alpha_i} \times Labor^{\beta_i} \times R\&D^{\gamma_i} \times Spillovers^{\delta_i} \times Advertising^{\phi_i}$$

The assumption that the elasticities differ across companies is easy to check, so we'll do that in a minute. Before doing so, however, I want to provide a precise definition of RQ:

RQ is the "company-specific output elasticity of R&D" (γ_i in Equation 3)

You can't just *compute* RQ. You have to "estimate" it statistically. Even if you never want to do this yourself, it's useful to see how the sausage is made so you understand what assumptions you're having to tolerate.

The first step in estimating the RQ for a company is to collect eight years of its financial data. For each year you need *revenues*, *capital* as net property, plant, and equipment, *labor* as full-time equivalent employees, *advertising*, and *R&D*. Next you combine the company's financial data with that for a large set of companies. I use all publicly traded companies in the United States. The data from these other companies is used to construct a secondary (and costless) input: company-specific *spillovers*. The spillover measure tries to capture technical knowledge that companies can adopt or imitate from leader companies. It's important to include this measure because it is the principal source of technical knowledge for small companies. As we learned in Chapter 2, most startups bootstrap their innovations from knowledge the founders gained at their prior employer. If we don't account for that, small company R&D looks more productive than it really is.

We match each year of inputs to the same year as output with the exception of R&D and spillovers. Here we make the assumption that R&D in one year isn't contributing to output until the following year, so we match the R&D expenditures from the year prior to the output year. Many people think this lag should be longer, but the statistical tests indicate across all companies the best-fitting lag is between one and two years. The reason the shorter lag makes sense is that while innovations may take many years from idea to market, very little of the spending occurs in the early years. The bulk of spending on R&D projects occurs very close to launch. So the "centroid" of spending appears to be one to two years before the company starts generating revenues from the innovation.

Once we have all this data, we perform regression analysis of Equation 3 to obtain the values of γ_i for each company. Each γ_i is the average "raw RQ" for that seven-year period (the eighth year of data is to accommodate lagged R&D). So when I refer to the company's fiscal year (FY) 2015 RQ, it is actually the average RQ over the period 2009 to 2015. To obtain an RQ trend, we merely drop out the oldest year, add a new year, and repeat the process.

To aid intuition I map the values for "raw RQ" onto the human IQ scale, so the average RQ across all companies in 2010 is 100, and the standard deviation across all companies is 15. This means that 66.7 percent of companies in 2010 have RQ values between 85 and 115. Like human genius, an exceptionally "smart" company is one whose RQ is above 130.

One aspect of RQ is that it is statistically *estimated*, rather than *computed*. This has two implications. First, RQ is backward looking—you only know what your RQ has been on average over those past seven periods. You never know exactly what it is this year (but you do know whether it is higher or lower than the prior year by comparing it to the prior RQ). Second, lots of factors affect a company's revenues year to year besides RQ. Accordingly, the RQ estimate at any time is only as good as our ability to control for those other factors.

In this sense, RQ is similar to a golf handicap. Golf handicaps are an effort to estimate a player's ability so that players of differing ability can be more evenly matched during competitions. On any given day, many factors affect a golfer's performance, such as weather, how recently the greens and fairways have been groomed, and the golfer's health and mental state. To compute your handicap, you first record the scores from the most recent rounds (minimum of 5 rounds to a maximum of 20 rounds). Thus a golf handicap, like RQ, is backward looking. Next, for each of the 5 to 20 rounds, you adjust for the difficulty of the course you played using its course rating and slope. The differences in course difficulty are equivalent to changes in the level of competition a company faces from period to period. Finally, you take the average (a simple form of statistical estimate) over all those adjusted scores to estimate true ability. Note that none of the adjustments account for the weather, how recently the course was groomed, or the player's health or state of mind. Thus even if a golfer's true ability is constant, his or her handicap will vary due to these other factors that don't go into the estimation. While estimating RQ is more complex than estimating a golfer's ability, in both cases, unless you can measure every factor affecting execution on a given day for a golfer, or financial period for a company, you can't *compute* the level of ability, you can only *estimate* it.

PROPERTIES OF RQ

Now that you understand how RQ is estimated, let's examine its properties.

RQ Differs Across Companies

The theoretical innovation behind RQ is the assumption that companies differ. So it's worth examining whether they do, and if so by how much. Figure 10-2 provides a histogram of RQs for all publicly traded companies conducting R&D. The figure indicates they do differ. The RQs for the highest companies are four times the average! One question economists often ask about these differences is whether they are just industry effects. "Isn't RQ just picking up differences in technological opportunity across industries?" If so, then understanding RQ is of limited value. Companies that know their industry RQ can make better investment decisions, but they'll be unable to exploit RQ differences or to improve their R&D effectiveness.

This turns out *not* to be the case. Figure 10-3 presents the range of RQs (min, max, and mean) for industries with the largest number of companies engaged in R&D. A few observations are worth noting. First, industries don't vary much in their mean RQ. The industry means range from a low of 86.6 for In Vitro Diagnostics (SIC 2835) to a high of 108.4 for Computer Storage Devices (SIC 3572). Thus industry means have a range of 21.8 RQ points. The second observation worth noting is there

FIGURE 10-2. **Organizational RQ (like individual IQ) is normally distributed**

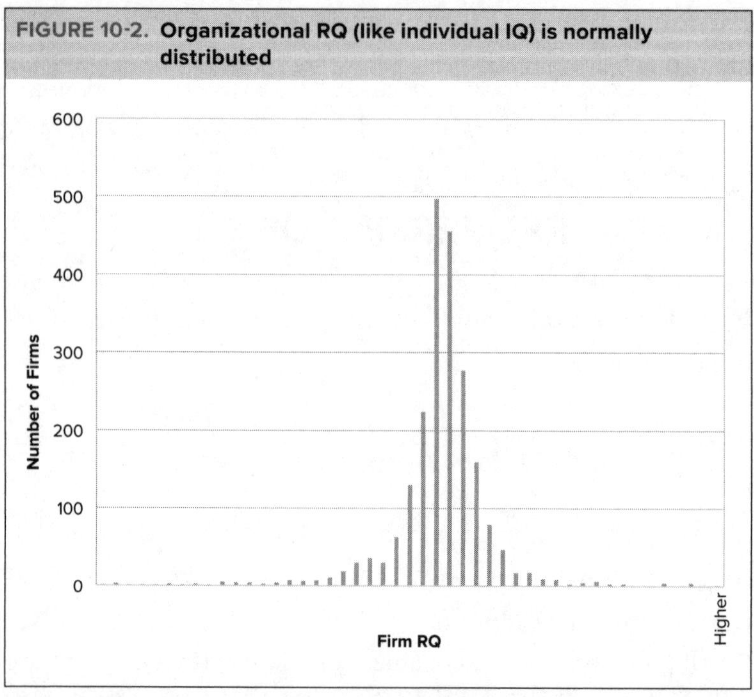

FIGURE 10-3. **RQ varies more within an industry than across industries**

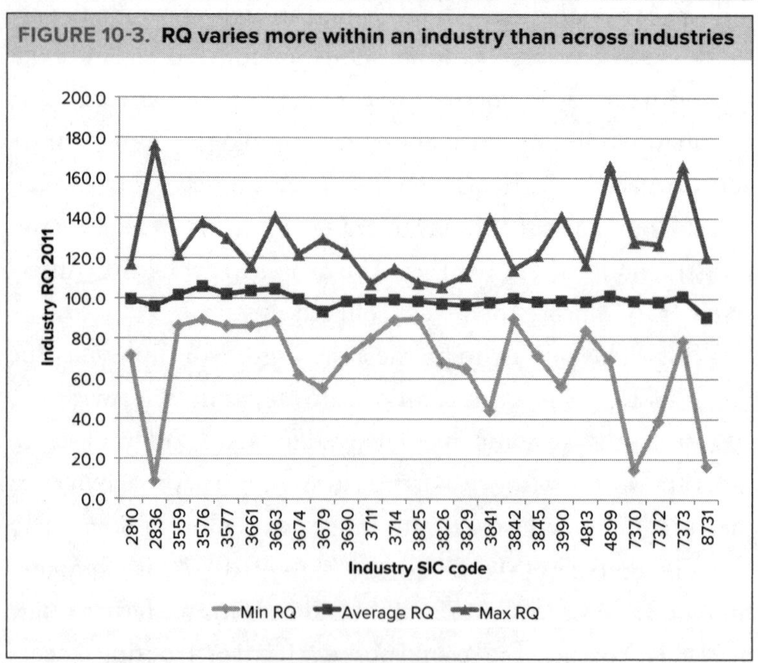

is greater variance of RQ within industries than across industries. In software (SIC 7372) for example, the minimum RQ is 38.2 and the maximum is 127.2 (a range of 89). There are however exceptions where a company's RQ seems to be defined by its industry. In optical instruments (SIC 3827), an industry with 10 public companies, the minimum RQ is 98.3, while the maximum is 101.8 (a range of only 3.5 RQ points).

Thus RQ differences are coming principally from *within* industry rather than *across* industry. This means companies can strategically exploit RQ!

RQ Is Universal

Because RQ is estimated entirely from financial data, it can be defined for *any* company doing R&D. Thus it is *universal*—one advantage it has over patents.

RQ Is Uniform

RQ is essentially a sophisticated ratio of outputs to inputs, where both are expressed in the same currency. This means RQ is unitless (dollars in the numerator cancel dollars in the denominator). This makes its interpretation *uniform* across companies regardless of their currency. Accordingly, RQs can be compared across companies within an industry and across industries, as well as across countries! This is another advantage RQ has over patents.

HOW DOES RQ COMPARE TO OTHER INNOVATION MEASURES?

Now that you've seen RQ, it is useful to compare it to measures you may be more familiar with. The most common measures companies use have been R&D turnover and patents. A third measure, used principally by economists, is total factor productivity (TFP). A final measure is Innovation Premium (IP), the basis of the Forbes Innovation 100 ranking that we discussed in Chapter 8. I will treat each of these measures in turn.

R&D Turnover

R&D turnover (revenues/R&D) is one of the most common measures of R&D productivity used by companies. This is a coarse measure of the revenue generating potential of R&D, because it doesn't take into account contributions from other inputs. R&D turnover is problematic as a measure because, as we discussed in Chapter 4, 78 percent of companies use its inverse (R&D/sales), commonly called R&D intensity, as a means to set their R&D budget. Thus R&D turnover may reflect "reverse causality"—sales drives the R&D budget, rather than R&D investment driving sales.

Patent Intensity

Patent counts, the number of patents granted to a company in a given year, and its various refinements (patent stock—the cumulative number of patents over a period, and patent citations—the number of citations of patents from a given year) are probably the most common measures of R&D output used by economists and management scholars. Patents become a measure of R&D productivity when expressed as patent intensity (patents/R&D expenditures).

One of the primary concerns with patents as a measure is they are not *universal*, meaning you can't use patents if you want to compare all companies. To provide a sense of how many companies patents exclude, we examined all public companies in the United States conducting R&D. We found that only 37 percent of them have any patents. Thus there is no way to measure the innovativeness of the other 63 percent of companies conducting R&D.

This begs the question, why do so few companies patent? One reason is that patents are costly both financially (the cost to file and defend) as well as competitively (they require disclosure of the fundamental knowledge underpinning the innovation). Accordingly, companies file patents only under certain circumstances. A Carnegie Mellon survey of innovating companies found that the primary use of patents was to prevent copying when innovations are easy to invent around, though companies also patent for strategic reasons, such as to block other companies' patents (82 percent of surveyed companies), to prevent lawsuits (59 percent), to use in negotiations (47 percent), or to enhance their reputation (48 percent).[2]

Not only do few companies patent their innovations, but those that do patent tend to avoid patenting their most valuable innovations. One intellectual property (IP) attorney inside a Fortune 50 company explained that the company only patents peripheral innovations. Innovations that are key to the company's strategy are protected via trade secret. Furthermore, no activity incorporating key technologies crosses company boundaries. Thus key technologies are never exposed to any form of open innovation, such as joint ventures or outsourcing. As he explained to me, core technologies outlast the life of patents, so patenting creates future competition.

There is evidence this company is the rule rather than the exception. In a recent paper examining patent trolls, David Abrams, Ufuk Akcigit, and Jillian Popadak found that the number of citations increases then decreases as the price paid for a patent increases[3] This is an indication that companies are less likely to patent their most valuable innovations. So not only do patent measures miss most companies, they miss the most important innovations inside companies.

A second issue with patents is they lack *uniformity*. All patents are not created equal. One study found that 10 percent of U.S. patents account for 81 to 85 percent of the economic value of all U.S. patents.

Total Factor Productivity (TFP)

Given these concerns with patents, economists have begun to use a new measure of companies' innovativeness called *total factor productivity* (*TFP*). TFP, like RQ, is derived from the company's production function, which as we saw in Equation 1 links a company's inputs to its outputs. Both TFP and RQ are efforts to capture the source of technological change (improvement) in the production function. However, they take different approaches. TFP assumes productivity improvement comes from an unknown source, whereas RQ assumes the improvement comes from the company's R&D. Accordingly, RQ is measured by teasing apart the productivity of each input (focusing on R&D), while TFP is measured as the remainder (the part of revenues that can't be explained by the contributions from measurable inputs). TFP is principally used in trade theory studies to examine the role of differences across companies in generating economic growth. These studies find that companies with higher TFP are more likely to export and grow.[4]

TFP shares two of the significant advantages RQ has over patent-based measures. First, it is universal—it can be estimated for any company. Second, TFP is unitless, so its interpretation is uniform across companies. However, while both TFP and RQ share the universality and uniformity advantage over patents, RQ is more meaningful to companies. This is because its foundational assumption (companies choose R&D to maximize company value) more closely matches what companies actually do than does the foundational assumption of TFP (innovation comes from undetermined causes). Further, because the innovation in TFP comes from undetermined causes, there is no way to define optimal R&D, nor to predict how any given level of R&D will affect future revenues, profits, and market value.

Holt's Innovation Premium (IP)

A final measure of innovativeness is Holt's IP, which we saw in Chapter 8. As described on the Forbes website, Innovation Premium (IP) is "the difference between [companies'] market capitalization and a net present value of cash flows from existing businesses. [This] is the bonus given by equity investors on the educated hunch that the company will continue to come up with profitable new growth." In essence, IP uses the combined wisdom of the market to measure how innovative a company is. One concern with this approach is that the market does a poor job of valuing innovation. This is why the RQ50 Index outperformed the market portfolio by 900 percent. The market only figures out the value of innovation after it shows up in profits and growth. In addition, growth can come from sources other than innovation. As an example, Monster Beverage is ranked 13 in the Forbes 100 (top IP companies). However, its website shows little evidence of product innovation. Its growth seems to come from a 5,000 percent increase in the energy drink category and expansion of distribution to meet that demand.

CORRELATIONS WITH OTHER INNOVATION MEASURES

So it's clear the various proxies for innovativeness measure different phenomena. Patents measure the counts of innovations, whereas TFP and IP measure unaccounted-for differences in companies' productivity and market value, respectively. In contrast, RQ measures R&D productivity, or returns to R&D investment. One question is whether all these measures, while computed differently, really capture the same underlying phenomenon. If so, then we should use whatever measure is most convenient.

We now explore whether that's true by looking at the correlation between each measure and RQ. R&D turnover really measures spending rather than output, and it is uncorrelated with RQ (correlation = 0.035). We already examined the correlation between IP and RQ in Chapter 8. We found that not only are IP and RQ uncorrelated, but also that the majority of companies ranked in Forbes 100 have below average RQ. We now examine the remaining measures.

Patent Intensity

We look next at patents (Figure 10-4). We use patent intensity (patents/R&D) because a raw count of patents really just measures company size. In contrast patent intensity captures the

FIGURE 10-4. **Correlation between patent intensity and RQ**

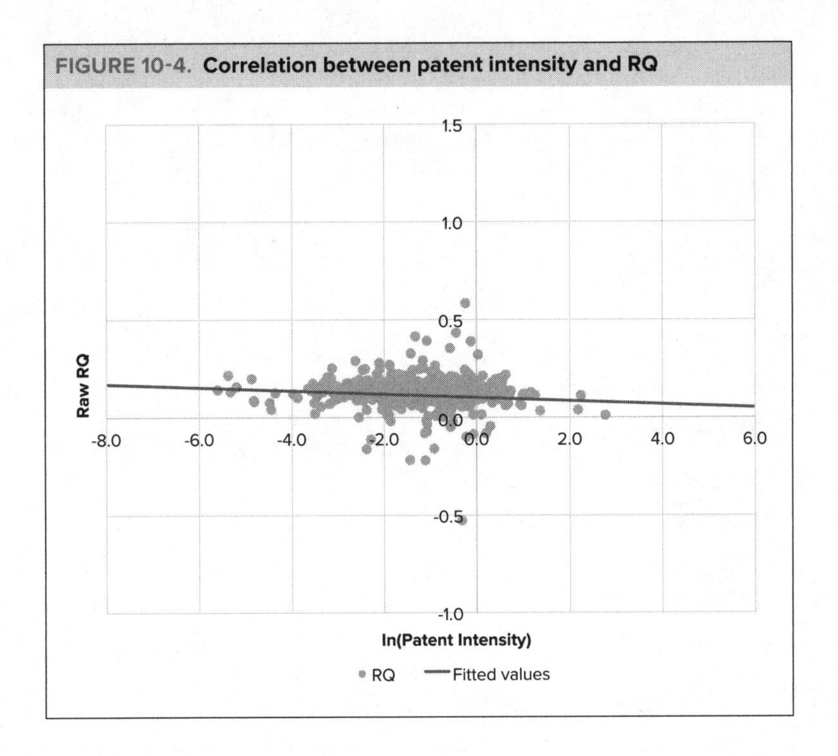

productivity of R&D in generating patents. What you can see in Figure 10-4 is that patent intensity and RQ are negatively correlated (correlation = –.143). Thus the more a company patents, the lower its RQ. While we know patents have problems of universality and uniformity, it's still a surprise they are negatively correlated with RQ. While we have no way of knowing for sure why this is true, one plausible explanation is that companies that are best able to exploit their R&D are more likely to use trade secrets rather than patents.

Total Factor Productivity (TFP)

Next we examine the correlation between TFP and RQ (Figure 10-5). Unlike patents, TFP is positively correlated with RQ

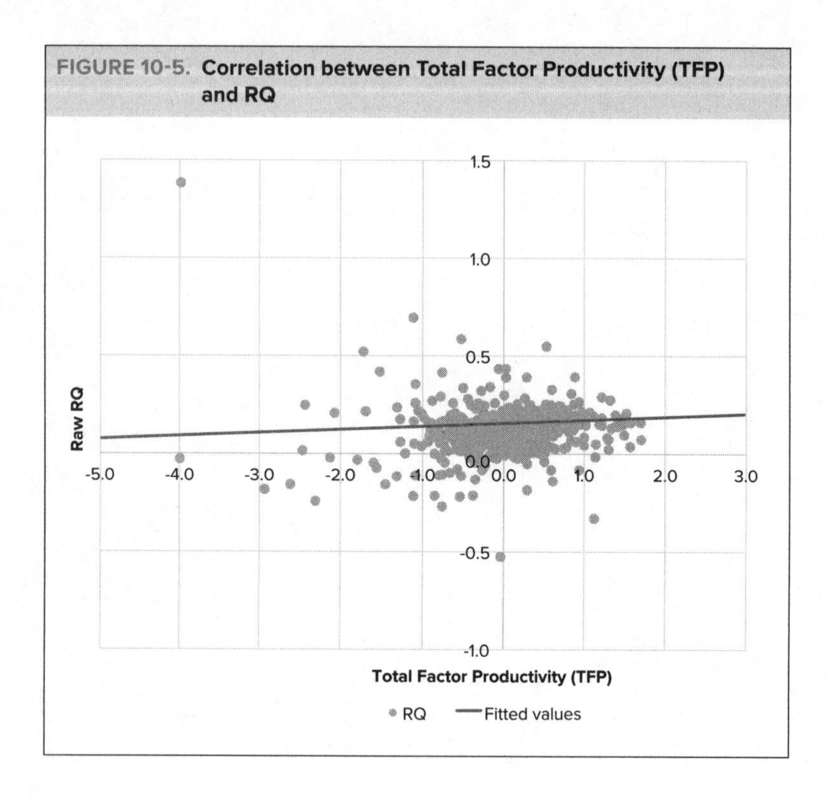

FIGURE 10-5. Correlation between Total Factor Productivity (TFP) and RQ

(correlation = .531). The likely reason for this is that R&D is one of the inputs that gets ignored when estimating the production function to generate TFP. What this means is that the effects of R&D will show up in the remainder (TFP).

THE REAL POWER OF RQ

We've seen that RQ has useful properties that you would like an innovation measure to have. It is normally distributed; it exhibits greater variance within than across industries; it is universal and uniform. But TFP exhibits these properties as well. As mentioned previously, the main advantage of RQ over TFP is that it assumes innovation is coming from R&D rather than some unidentified factor. This allows us to use RQ to define the optimal level of R&D, and to predict the future revenues, profits, and market value associated with that level of R&D.

To see how to link RQ to these financial outcomes, let's create a fictitious company whose "raw RQ", γ, is 0.10, revenues are $10 billion, R&D is $400 million, gross margin is 30 percent, net income (NI) is $900 million, and market value is $18 billion.

Revenues. The most straightforward prediction is revenues. RQ is defined as the output elasticity of R&D; thus if we define output as revenues, a 1 percent increase in the company's R&D increases its revenues γ percent. In our example, let's increase R&D by $40 million (representing a 10 percent increase). Then revenues increase by 10 × 0.01 = 1%. Since current revenues are $10 billion, revenues will increase $100 million from the additional $40 million in R&D.

Profits. Once we know the change in revenues, we merely need to subtract the change in costs to obtain profits. Assuming that the company still retains the 30 percent gross margin, then the new profits are merely the gross margin times the new revenues (0.3 × $100 million = $30 million) minus the additional R&D ($40 million). In this case, because the additional gross margin is $30 million, while the additional R&D is $40 million, profits would actually fall $10 million. This tells us that $440 million is above the optimal level of R&D for this company.

Market value. Finally, if we assume the company retains its P/E ratio of 20 ($18 billion/$900 million), then the $10 million decrease in profits will generate a $200 million decrease in market value.

DETERMINING OPTIMAL R&D

One final important relationship that can be derived from RQ is the optimal level of R&D spending. We discussed optimal R&D in Chapter 4, but here we tell you how to calculate it. You could certainly find it by trial and error by entering the equation for profits and the company's starting values for RQ (0.10) and gross margin (30 percent) into a spreadsheet, then testing various levels of R&D to see when profits are highest. However, there is a much a quicker way if you're comfortable with calculus and know all the input levels and their elasticities.

The starting point is building an equation for profits from the production function in Equation 2.

(4)
$$Profits = (Capital^{\alpha} \times Labor^{\beta} \times R\&D^{\gamma} \times Spillovers^{\delta} \times Advertising^{\varphi}) - Costs$$

Next we want to unpack costs. We'll make the same simplifying assumption we did above that the contribution margin stays the same, so the only other cost to consider is R&D. If so, then profits are defined as:

(5)
$$Profits = (gross\ margin) \times (Capital^{\alpha} \times Labor^{\beta} \times R\&D^{\gamma} \times Spillovers^{\delta} \times Advertising^{\varphi}) - R\&D$$

The next principle to remember, from microeconomics, is that the way to find the profit-maximizing level of an input (assuming everything else is fixed) is to take the derivative of profits with respect to that input and set it equal to zero. So to find the optimal value of R&D, which we will call R* (or R-star), we take the derivative of the profit equation with respect to R&D. I've worked it out for you:[5]

(6)
$$R_i^* = (\alpha\gamma_i Capital^{\alpha} Labor^{\beta} Spillovers^{\delta} Advertising^{\varphi} e^{\beta 0})^{\frac{1}{1-\gamma_i}}$$

THE RELIABILITY OF RQ

The relationships in the equations for revenues, profits, and optimal R&D are simplified expressions of the relationships in company-level models of endogenous growth theory that we discussed in Chapter 4.[6] This is the theory linking R&D investment to growth. What's interesting is that for at least a century companies have done R&D to generate growth, but the theory to characterize how that worked wasn't developed until recently. The endogenous growth models treat R&D as investments made to maximize company value. The role of R&D in these models is incremental contribution to a knowledge stock that enhances the productivity of other inputs. The ability of companies to grow the knowledge stock and accordingly their revenues is conditioned by their R&D productivity. The models generate a number of testable propositions regarding R&D productivity. In particular, companies with higher R&D productivity are expected to have higher optimal R&D investment, higher profits, higher growth, and higher market value.

There are two important features to having a measure derived from theory. First, you can test propositions from the model to validate the theory, and second, you can validate measures by seeing if they match predictions from the model. A colleague, Carl Vieregger, and I did that for the set of publicly traded companies conducting R&D from 1965 to 2011.[7] We tested the three main measures of company innovativeness discussed earlier: patent intensity, total factor productivity (TFP), and RQ.

When we did that, we found that all three propositions from the model were supported when using RQ as the measure of R&D productivity. R&D investment, market value, and growth all increase with RQ. Thus the theory itself appears valid, and in addition the RQ measure appears to capture the model's construct for R&D productivity.

However, we also found that neither of the other measures (patent intensity nor TFP) was consistent with all three propositions. While R&D investment and market value both increase with a company's TFP, growth actually decreases with TFP. Interestingly, R&D investment, market value, and growth all *decrease* with patent intensity. Accordingly, neither patents nor TFP is a reliable measure of R&D productivity.

What's important about this set of tests for academics is they demonstrate RQ is a valid measure of R&D productivity. What's important about the tests for companies is they tell companies they can reliably use RQ to set R&D investment levels, and to predict how much that investment will increase their revenues, profits, market value, and growth.

OBTAINING YOUR COMPANY'S RQ

Because the RQ measure is rooted in academic research, the estimation methodology and the validation tests are all in the public domain. If you would like to develop internal capability to generate RQ on a regular basis, you need to hire someone with statistical expertise in multilevel models, subscribe to a database of 10-K information for all publicly traded companies in the United States, and obtain a license for multilevel statistical modeling.

However, there is an easier and less expensive option. If you work for a publicly traded company, you can obtain your company's RQ for free at amkANALYTICS.com. If you work for a privately held company, you won't be able to obtain a free report because your financial data isn't included in 10-K databases.

However, you can obtain a reader discount for custom (and more detailed) analysis of your RQ that is generated from secure entry of your financial data. In addition to generating your RQ, the analysis will allow you to simulate changes to your R&D and forecast the impact of those changes on your revenues, profits, and market value.

SUMMARY

RQ is the most powerful measure you can construct for a company's innovativeness. It is derived from economic theory, and more important, its behavior matches propositions from economic theory when tested over 47 years of data: RQ predicts R&D investment, market value, and revenue growth. Thus it is *valid*.

However, RQ also has properties that make it useful to companies. The data to estimate RQ is *easy to collect*. In fact, for public companies, it's data they already collect for their SEC Form 10-K. RQ is *uniform*—it is essentially a ratio of output to inputs, so it can be compared across business units. RQ *can't be gamed* to make a given group look good because it involves fairly sophisticated regression that takes into account contributions from other inputs. Thus the only way for a group to look good is to *be* good. Perhaps most important, RQ is meaningful to shareholders. It's the single most significant predictor of monthly stock returns over the past 47 years.

Given all that, RQ is a valuable tool for (1) justifying R&D investment to CEOs, the board and investors, (2) improving the efficiency of R&D, and (3) estimating the value of R&D investment to future growth.

APPENDIX

Now that we understand how important RQ is, it's useful to see what high RQ looks like. The RQ50 are the publicly traded companies whose R&D investment creates the greatest value for their shareholders. To create the list in Table A-1, I compiled the set of all publicly held U.S. companies investing $100 million or more in R&D FY2015, ranked them according to their RQ, and took the top 50.

Note: the RQ50 portfolio discussed in Chapter 8 differs from this list in that it has no R&D spending threshold and it includes foreign firms that trade on the U.S. exchanges.

TABLE A-1. Fiscal Year 2015 RQ50

	COMPANY	SALES ($ MILLION)	R&D ($ MILLION)	INDUSTRY
1	MCKESSON CORP	190884	392	Wholesale Drugs
2	MEDIVATION INC	943	190	Biological Products
3	MEDICINES CO	309	159	Pharmaceuticals
4	SANDISK CORP	5565	852	Computer Storage Devices
5	SYNAPTICS INC	1703	193	Semiconductors
6	TIVO CORP	490	102	Cable Services
7	GILEAD SCIENCES INC	32639	2854	Biological Products
8	HEARTWARE INTERNATIONAL INC	277	122	Medical Equipment
9	ADVANCED MICRO DEVICES	3991	1072	Semiconductors
10	ALEXION PHARMACEUTICALS INC	2604	514	Biological Products
11	SILICON LABORATORIES INC	645	173	Semiconductors
12	NEWMARKET CORP	2141	139	Industrial Chemicals
13	ARRIS INTERNATIONAL PLC	4798	557	Communications Equipment
14	LYONDELLBASELL INDUSTRIES NV	32735	127	Plastics Materials
15	CELGENE CORP	9256	2431	Pharmaceuticals
16	INTUITIVE SURGICAL INC	2384	178	Medical Equipment
17	INFINITY PHARMACEUTICALS INC	109	144	Pharmaceuticals
18	LAM RESEARCH CORP	5259	716	Machinery
19	CIRRUS LOGIC INC	1169	198	Semiconductors
20	AMAZON.COM INC	107006	9275	Mail Order
21	UNITED THERAPEUTICS CORP	1466	243	Pharmaceuticals
22	HASBRO INC	4448	223	Toys
23	ALTRIA GROUP INC	18854	167	Cigarettes
24	TAKE-TWO INTERACTIVE SOFTWARE	1414	115	Software

	COMPANY	SALES ($ MILLION)	R&D ($ MILLION)	INDUSTRY
25	ACTIVISION BLIZZARD INC	4664	571	Software
26	EXXON MOBIL CORP	236810	971	Petroleum
27	XILINX INC	2214	526	Semiconductors
28	FIRST SOLAR INC	3579	144	Semiconductors
29	NETFLIX INC	6780	472	Video Rental
30	BIOGEN INC	10764	1910	Biological Products
31	QLOGIC CORP	459	144	Computer Equipment
32	ROVI CORP	526	109	Patent Owner
33	F5 NETWORKS INC	1920	264	System Design
34	CIENA CORP	2446	401	Communications Equipment
35	IAC/INTERACTIVECORP	3231	161	Computer Services
36	QUALCOMM INC	25281	5477	Semiconductors
37	ALPHABET INC	74989	9832	Computer Services
38	EBAY INC	8592	2000	Computer Services
39	AMGEN INC	21662	4297	Biological Products
40	POLARIS INDUSTRIES INC	4719	148	Transportation Equipment
41	APPLE INC	233715	6041	Communications Equipment
42	ZIMMER BIOMET HOLDINGS INC	5998	188	Medical Equipment
43	CONOCOPHILLIPS	29564	263	Petroleum
44	IONIS PHARMACEUTICALS INC	284	156	Pharmaceuticals
45	APPLIED MATERIALS INC	9659	1428	Machinery
46	DOW CHEMICAL	48778	1647	Plastics Materials
47	HUNTSMAN CORP	10299	158	Industrial Chemicals
48	VERIFONE SYSTEMS INC	2000	204	Accounting Machines
49	DOLBY LABORATORIES INC	971	183	Patent Owner
50	MICROSEMI CORP	1246	192	Semiconductors

The companies represent a broad swath of the economy (Figure A-1). The biggest representation (16 percent) comes from semiconductors, followed by biotech (10 percent) and pharmaceuticals (10 percent), but fully 18 percent of the RQ50 are the only companies in their industry to make the cut. So it's not the case that these companies are all riding the same wave of opportunity. The RQ50 is fairly stable—on average 70 percent of companies who make the list one year appear the following year. Perhaps the most surprising revelation from the list, however, is that you probably don't recognize most of the companies. There is very little overlap between the RQ50 and other innovation rankings, such as those of *Forbes*, *Fast Company*, Boston Consulting Group, and Strategy& (formerly Booz). In fact, the only RQ50 company to appear consistently in other rankings is Amazon. Why the discrepancy? In the case

FIGURE A-1. Industries in the RQ50

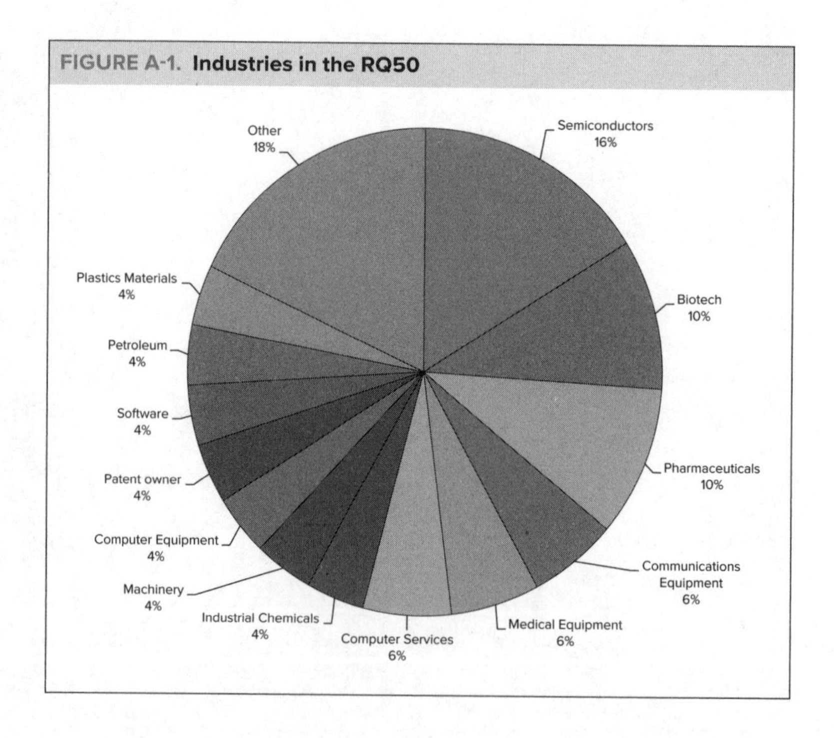

of BCG and *Fast Company*, the rankings are based on opinion. Accordingly, a high ranking requires innovations so visible that editors and executives outside the industry notice them. That is not where opportunity lies. In contrast, the RQ50 captures results before they cross editors' radar.

It is not clear from looking at the list what the RQ50 have in common that makes them high RQ. However, it *is* clear that it doesn't line up with conventional views of what innovativeness looks like. That's why RQ is so important—it captures the *reality* of what makes companies innovative, rather than the *perception* of what makes them innovative. You may be able to deduce some insights into what characteristics set the RQ50 apart by reviewing the company profiles that follow in greater detail.

1 MCKESSON CORP

MCK

FY 2015 RANK	FY 2014 RANK	FY 2015 SALES ($ MILLION)	FY 2015 R&D INVESTMENT ($ MILLION)	FY 2015 GROSS MARGIN
1	2	190,884.0	392.0	6.1%

Scope of R&D:
Development expenditures are primarily incurred by our Technology Solutions segment. Our Technology Solutions segment's product development efforts apply computer technology and installation methodologies to specific information processing needs of hospitals and other customers.

Chief Executive Officer . . . . John H. Hammergren
Chief Technology Officer . . . Kathy McElligott
Industry . . . . Wholesale Drugs
Headquarters. . . . . San Francisco, CA
Website . . . . www.mckesson.com
Years in RQ50 Index 12

---

2 MEDIVATION INC

MDVN

FY 2015 RANK	FY 2014 RANK	FY 2015 SALES ($ MILLION)	FY 2015 R&D INVESTMENT ($ MILLION)	FY 2015 GROSS MARGIN
2	1	943.3	189.6	46.7%

Scope of R&D:
We plan to continue to invest in our ongoing clinical studies to cover the continuum of prostate cancer care, advance our development of enzalutamide for the treatment of advanced breast cancer, fully enroll our gBRCA mutated advanced breast cancer clinical trial for MDV3800, advance MDV9300 in DLBCL and prepare for other hematologic indications such as multiple myeloma, and further the development and growth of our internal pipeline. Furthermore, we also plan to initiate startup activities for a number of phase 2 and potentially pivotal clinical trials to develop MDV3800 beyond gBRCA mutations.

Chief Executive Officer . . . . David Hung, MD
Chief Medical Officer . . . . . Mohammad Hirmand, MD
Industry . . . . Biological Products
Headquarters. . . . . San Francisco, CA
Website . . . . www.medivation.com
Years in RQ50 Index 2

---

3 MEDICINES CO MDCO

FY 2015 RANK	FY 2014 RANK	FY 2015 SALES ($ MILLION)	FY 2015 R&D INVESTMENT ($ MILLION)	FY 2015 GROSS MARGIN
3	3	309.0	159.2	68.3%

Scope of R&D:
The company uses a proprietary innovation framework and system to identify market opportunities and needs. It then discovers, licenses, and develops biopharmaceutical technologies, achieves regulatory approvals and reimbursement, and manages market place change.

Chief Executive Officer . . . . Clive A. Meanwell, MD, PhD
Industry . . . . Pharmaceuticals
Headquarters. . . . . Parsippany, NJ
Website . . . . www.themedicinescompany.com
Years in RQ50 Index 6

4 SANDISK CORP SNDK

FY 2015 RANK	FY 2014 RANK	FY 2015 SALES ($ MILLION)	FY 2015 R&D INVESTMENT ($ MILLION)	FY 2015 GROSS MARGIN
4	5	5,564.9	852.3	48.0%

Scope of R&D:
Since our inception, we have focused our research, development and standardization efforts on developing highly reliable, high-performance, cost-effective NAND flash storage products in many form factors to address a variety of markets. Our research and development, or R&D, efforts are designed to help ensure the creation of fully-integrated, broadly interoperable products that are compatible with both existing and newly developed system platforms.

Chief Executive Officer . . . . Sanjay Mehrotra
Senior Vice President and Chief Technology Officer. . Kevin Conley
Industry . . . . Computer Storage Devices
Headquarters. . . . . Milpitas, CA
Website . . . . www.sandisk.com
Years in RQ50 Index 11

5 SYNAPTICS INC

SYNA

FY 2015 RANK	FY 2014 RANK	FY 2015 SALES ($ MILLION)	FY 2015 R&D INVESTMENT ($ MILLION)	FY 2015 GROSS MARGIN
5	11	1,703.0	192.7	39.7%

Scope of R&D:
We conduct ongoing research and development programs that focus on advancing our existing interface technologies, improving our current product solutions, developing new products, improving design and manufacturing processes, enhancing the quality and performance of our product solutions, and expanding our technologies to serve new markets. Our goal is to provide our customers with innovative solutions that address their needs and improve their competitive positions. Our long-term vision is to offer human interface semiconductor product solutions, such as touch, fingerprint, handwriting, vision, voice capabilities, and other biometrics that can be readily incorporated into various electronic devices.

Chief Executive Officer . . . . Rick Bergman
Senior Vice President, Corporate Research and CTO . . Patrick Worfolk
Industry . . . . Semiconductors
Headquarters. . . . . San Jose, CA
Website . . . . www.synaptics.com
Years in RQ50 Index 4

6 TIVO CORP

TIVO

FY 2015 RANK	FY 2014 RANK	FY 2015 SALES ($ MILLION)	FY 2015 R&D INVESTMENT ($ MILLION)	FY 2015 GROSS MARGIN
6	10	489.6	102.2	63.4%

Scope of R&D:
Our research and development efforts are focused on designing and developing the elements necessary to enable the TiVo service. These activities include hardware and software development.

Interim CEO. . . . . Naveen Chopra
Industry . . . . Cable Services
Headquarters. . . . . San Jose, CA
Website . . . . www.tivo.com
Years in RQ50 Index 4

7 GILEAD SCIENCES INC GILD

FY 2015 RANK	FY 2014 RANK	FY 2015 SALES ($ MILLION)	FY 2015 R&D INVESTMENT ($ MILLION)	FY 2015 GROSS MARGIN
7	13	32,639.0	2,854.0	90.8%

Scope of R&D:
Our R&D philosophy and strategy is to develop best-in-class drugs that improve safety or efficacy for unmet medical needs. Our product development efforts cover a wide range of medical conditions, including HIV/AIDS and liver diseases such as HBV and HCV, inflammation/oncology and serious cardiovascular and respiratory conditions.

Chief Executive Officer . . . . John F. Milligan, PhD
Executive Vice President, Research and Development and Chief Scientific Officer . . . . . Norbert W. Bischofberger, PhD
Industry . . . . Biological Products
Headquarters . . . . Foster City, CA
Website . . . . www.gilead.com
Years in RQ50 Index 2

---

8 HEARTWARE INTERNATIONAL INC HTWR

FY 2015 RANK	FY 2014 RANK	FY 2015 SALES ($ MILLION)	FY 2015 R&D INVESTMENT ($ MILLION)	FY 2015 GROSS MARGIN
8	8	276.8	122.4	68.9%

Scope of R&D:
Research and development costs include activities related to the research, development, design, testing, and manufacturing of prototypes of our products as well as costs associated with certain clinical and regulatory activities. We expect our research and development expenses to continue to increase as we implement enhancements to the HVAD System, continue to develop our MVAD System and CircuLite System, enhance our peripheral product offerings, conduct additional pre-approval and post-approval clinical trials and hire additional employees.

Chief Executive Officer . . . . Doug Godshall
Senior Vice President, Research & Development and Quality . . . . Mark Strong
Industry . . . . Medical Equipment
Headquarters . . . . Framingham, MA
Website . . . . www.heartware.com
Years in RQ50 Index 2

---

9 ADVANCED MICRO DEVICES AMD

FY 2015 RANK	FY 2014 RANK	FY 2015 SALES ($ MILLION)	FY 2015 R&D INVESTMENT ($ MILLION)	FY 2015 GROSS MARGIN
9	12	3,991.0	1,072.0	30.2%

Scope of R&D:
We focus our research and development activities on improving product performance and enhancing product design. Our main area of focus is on delivering the next generation of CPU and GPU IP, and designing that IP into our SoCs for our next generation of products, with, in each case, improved system performance and performance-per-watt characteristics. For example, we are focusing on improving the battery life of our microprocessors and APU products for notebooks and the power efficiency of our microprocessors for servers. We are also focusing on delivering a range of low-power integrated platforms to serve key markets, including commercial clients, mobile computing and gaming and media computing. We believe that these platforms will bring customers increased performance and energy efficiency. We also work with industry leaders on process technology, software and other functional intellectual property and we work with others in the industry and industry consortia to conduct early stage research and development.

Chief Executive Officer . . . . Dr. Lisa Su
Chief Technology Officer and Senior Vice President. . . Mark Papermaster
Industry . . . . Semiconductors
Headquarters. . . . . Sunnyvale, CA
Website . . . . www.amd.com
Years in RQ50 Index 8

---

10 ALEXION PHARMACEUTICALS INC ALXN

FY 2015 RANK	FY 2014 RANK	FY 2015 SALES ($ MILLION)	FY 2015 R&D INVESTMENT ($ MILLION)	FY 2015 GROSS MARGIN
10	14	2,604.0	513.8	92.7%

Scope of R&D:
Our research and development expense includes personnel, facility and external costs associated with the research and development of our product candidates, as well as product development costs. We group our research and development expenses into two major categories: external direct expenses and all other research and development (R&D) expenses. External direct expenses are comprised of costs paid to outside parties for clinical development, product development and discovery research, as well as costs associated with strategic licensing agreements we have entered into with third parties. Clinical development costs are comprised of costs to conduct and manage clinical trials related to eculizumab and other product candidates. Product development costs are those incurred in performing duties related to manufacturing development and regulatory functions, including

manufacturing of material for clinical and research activities. Discovery research costs are incurred in conducting laboratory studies and performing preclinical research for other uses of our products and other product candidates. Licensing agreement costs include upfront and milestone payments made in connection with strategic licensing arrangements we have entered into with third parties.

Chief Executive Officer . . . . David Hallal
Executive Vice President, Global Head of Research & Development . . . . Martin Mackay, PhD
Industry . . . . Biological Products
Headquarters. New Haven, CT
Website . . . . www. alexion.com
Years in RQ50 Index 2

---

11 SILICON LABORATORIES INC SLAB

FY 2015 RANK	FY 2014 RANK	FY 2015 SALES ($ MILLION)	FY 2015 R&D INVESTMENT ($ MILLION)	FY 2015 GROSS MARGIN
11	7	644.8	173.0	61.7%

Scope of R&D:
Silicon Labs is a leading provider of silicon, software and solutions for the Internet of Things, Internet infrastructure, industrial automation, consumer and automotive markets. Through our R&D efforts, we leverage analog/mixed-signal engineering to create highly integrated IC products characterized by lower costs, smaller die sizes, reduced power demands and enhanced performance. We specialize in creating general-purpose hardware and software platform architectures that can be used across multiple applications and market segments.

Chief Executive Officer . . . . Tyson Tuttle
Chief Technology Officer . . . Alessandro Piovaccari
Industry . . . . Semiconductors
Headquarters. . . . . Austin, TX
Website . . . . www.silabs.com
Years in RQ50 Index 9

---

12 NEWMARKET CORP NEU

FY 2015 RANK	FY 2014 RANK	FY 2015 SALES ($ MILLION)	FY 2015 R&D INVESTMENT ($ MILLION)	FY 2015 GROSS MARGIN
12	15	2,140.8	139.2	33.6%

Scope of R&D:
The petroleum additives industry is subject to periodic technological change, changes in performance standards, and ongoing product improvements. Further, technological changes in some or all of our customers' products or processes may make our products obsolete. As a result, the life cycle of our products is often hard to predict. In order to maintain our profits and remain competitive, we must effectively respond to technological changes in our industry and successfully develop, manufacture, and market new or improved products in a cost-effective and timely manner.

Chief Executive Officer . . . . Thomas Gottwald
Industry . . . . Industrial Chemicals
Headquarters. . . . . Richmond, VA
Website . . . . www.newmarket.com
Years in RQ50 Index 4

---

13 ARRIS INTERNATIONAL PLC ARRS

FY 2015 RANK	FY 2014 RANK	FY 2015 SALES ($ MILLION)	FY 2015 R&D INVESTMENT ($ MILLION)	FY 2015 GROSS MARGIN
13	9	4,798.3	556.6	31.2%

Scope of R&D:
We continue to innovate in anticipation of both our customers' needs and developing industry trends, including: 1) solving the complexity of delivering content to the growing number of connected devices, 2) anticipating demand for more personalized, relevant, and mobile experiences, 3) monetizing future services like UHD, IoT, Gigabit Wi-Fi and multiscreen, and transitioning to all-IP networks, 4) employing state-of-the-art computing and packet processing technologies to solve the last-mile bottleneck, 5) the enablement of complex function to be performed in the "cloud".

Chief Executive Officer . . . . Bruce McClelland
Senior Vice President, Chief Information Officer. . . Phil Baldock
Industry . . . . Communications Equipment
Headquarters. . . . . Suwanee, GA
Website . . . . www.arris.com
Years in RQ50 Index 7

---

14 LYONDELLBASELL INDUSTRIES NV LYB

FY 2015 RANK	FY 2014 RANK	FY 2015 SALES ($ MILLION)	FY 2015 R&D INVESTMENT ($ MILLION)	FY 2015 GROSS MARGIN
14		32,735.0	127.0	24.8%

Scope of R&D:
Our research and development ("R&D") activities are designed to improve our existing products and processes, and discover and commercialize new materials, catalysts and processes. These activities focus on product and application development, process development, catalyst development and fundamental polyolefin focused research.

Chief Executive Officer . . . . Bhavesh V. (Bob) Patel
Chief Information Officer . . . Marc Franciosa
Industry . . . . Plastics Materials
Headquarters. . . . . Houston, TX
Website . . . . www.lyondellbasell.com
Years in RQ50 Index 1

---

15 CELGENE CORP CELG

FY 2015 RANK	FY 2014 RANK	FY 2015 SALES ($ MILLION)	FY 2015 R&D INVESTMENT ($ MILLION)	FY 2015 GROSS MARGIN
15	25	9,256.0	2,430.6	96.8%

Scope of R&D:
We continue to invest substantially in research and development in support of multiple ongoing proprietary clinical development programs which support our existing products and pipeline of new drug candidates. Our clinical trial activity includes trials across the disease areas of hematology, oncology, and inflammation and immunology.

Chief Executive Officer . . . . .Mark J. Alles
President of Research and Early Development. . .Rupert Vessey, PhD
Industry . . . . .Pharmaceuticals
Headquarters. . . . . .Summit, NJ
Website . . . . .www.celgene.com
Years in RQ50 Index. 9

---

16 INTUITIVE SURGICAL INC ISRG

FY 2015 RANK	FY 2014 RANK	FY 2015 SALES ($ MILLION)	FY 2015 R&D INVESTMENT ($ MILLION)	FY 2015 GROSS MARGIN
16	22	2,384.4	178.0	69.8%

Scope of R&D:
We focus our research and development efforts on providing our customers with new products and product improvements that enable them to perform MIS procedures with less difficulty.

Chief Executive Officer . . . . Gary S. Guthart, PhD
Senior Vice President, Product Development . . Brian Miller, PhD
Industry . . . . Medical Equipment
Headquarters. . . . . Sunnyvale, CA
Website . . . . www.intuitivesurgical.com
Years in RQ50 Index 5

17 INFINITY PHARMACEUTICALS INC INFI

FY 2015 RANK	FY 2014 RANK	FY 2015 SALES ($ MILLION)	FY 2015 R&D INVESTMENT ($ MILLION)	FY 2015 GROSS MARGIN
17		109.1	143.6	negative

Scope of R&D:
Our research and development group is focusing on drug discovery, preclinical research, translational medicine, clinical trials and manufacturing technologies.

Chief Executive Officer . . . . Adelene Perkins
President, Research and Development . . Julian Adams, PhD
Industry . . . . Pharmaceuticals
Headquarters. . . . . Cambridge, MA
Website . . . . www.infi.com
Years in RQ50 Index 1

18 LAM RESEARCH CORP LRCX

FY 2015 RANK	FY 2014 RANK	FY 2015 SALES ($ MILLION)	FY 2015 R&D INVESTMENT ($ MILLION)	FY 2015 GROSS MARGIN
18	21	5,259.3	716.5	49.0%

Scope of R&D:
We continued to make significant R&D investments focused on leading-edge deposition, plasma etch, single-wafer clean and other semiconductor manufacturing requirements.

Chief Executive Officer . . . . Martin Anstice
Senior Vice President and Chief Technology Officer . . Dave Hemker
Industry . . . . Machinery
Headquarters. . . . . Fremont, CA
Website . . . . www.lamresearch.com
Years in RQ50 Index 16

19 CIRRUS LOGIC INC CRUS

FY 2015 RANK	FY 2014 RANK	FY 2015 SALES ($ MILLION)	FY 2015 R&D INVESTMENT ($ MILLION)	FY 2015 GROSS MARGIN
19		1,169.3	197.9	52.4%

Scope of R&D:
We concentrate our research and development efforts on the design and development of new products for each of our principal markets. We also fund certain advanced-process technology development, as well as other emerging product opportunities.

Chief Executive Officer . . . . Jason Rhode
Industry . . . . Semiconductors
Headquarters. . . . . Austin, TX
Website . . . . www.cirrus.com
Years in RQ50 Index 9

20 AMAZON.COM INC AMZN

FY 2015 RANK	FY 2014 RANK	FY 2015 SALES ($ MILLION)	FY 2015 R&D INVESTMENT ($ MILLION)	FY 2015 GROSS MARGIN
20	6	107,006.0	9,275.0	37.6%

Scope of R&D:
We seek to invest efficiently in several areas of technology and content so we may continue to enhance the customer experience and improve our process efficiency through rapid technology developments while operating at an ever increasing scale. Our technology and content investment and capital spending projects often support a variety of product and service offerings due to geographic expansion and the cross-functionality of our systems and operations.

Chief Executive Officer . . . . Jeff Bezos
Chief Technology Officer . . . Werner Hans Peter Vogels, PhD
Industry . . . . Mail Order
Headquarters. . . . . Seattle, WA
Website . . . . www.amazon.com
Years in RQ50 Index 15

21 UNITED THERAPEUTICS CORP UTHR

FY 2015 RANK	FY 2014 RANK	FY 2015 SALES ($ MILLION)	FY 2015 R&D INVESTMENT ($ MILLION)	FY 2015 GROSS MARGIN
21	28	1,465.8	242.5	97.5%

Scope of R&D:
We are engaged in a number of research and development activities in xenotransplantation, regenerative medicine and ex-vivo lung perfusion, all of which are intended to increase the supply of transplantable organs and tissues. These activities are principally focused on lungs, but are also being applied to other organs such as hearts and kidneys.

Chief Executive Officer . . . . Martine Rothblatt
Industry . . . . Pharmaceuticals
Headquarters. . . . . Silver Spring, MD
Website . . . . www.unither.com
Years in RQ50 Index 3

22 HASBRO INC HAS

FY 2015 RANK	FY 2014 RANK	FY 2015 SALES ($ MILLION)	FY 2015 R&D INVESTMENT ($ MILLION)	FY 2015 GROSS MARGIN
22	26	4,447.5	222.6	55.3%

Scope of R&D:
Our success is dependent on continuous innovation in our branded-play and entertainment offerings and requires continued development of new brands and products alongside the redesign of existing products to drive consumer interest and market acceptance. Our toy and game products are developed by a global development function.

Chief Executive Officer . . . . Brian D. Goldner
Industry . . . . Toys
Headquarters. . . . . Pawtucket, RI
Website . . . . www.hasbro.com
Years in RQ50 Index 23

23 ALTRIA GROUP INC MO

FY 2015 RANK	FY 2014 RANK	FY 2015 SALES ($ MILLION)	FY 2015 R&D INVESTMENT ($ MILLION)	FY 2015 GROSS MARGIN
23		18,854.0	167.0	60.0%

Scope of R&D:
Altria Group works to meet evolving adult tobacco consumer preferences over time by developing, manufacturing, marketing and distributing products.

Chief Executive Officer . . . . Martin Barrington
Chief Innovation Officer. . . . James E. Dillard III
Industry . . . . Cigarettes
Headquarters. . . . . Richmond, VA
Website . . . . www.altria.com
Years in RQ50 Index 10

24 TAKE-TWO INTERACTIVE SOFTWARE TTWO

FY 2015 RANK	FY 2014 RANK	FY 2015 SALES ($ MILLION)	FY 2015 R&D INVESTMENT ($ MILLION)	FY 2015 GROSS MARGIN
24		1,413.7	115.0	42.4%

Scope of R&D:
We had a research and development staff of 2,179 employees with the technical capabilities to develop software titles for all major consoles, handheld hardware platforms and PCs in multiple languages and territories.

Chief Executive Officer . . . . Strauss Zelnick
Industry . . . . Software
Headquarters. . . . . New York, NY
Website www.take2games.com
Years in RQ50 Index 2

---

25 ACTIVISION BLIZZARD INC ATVI

FY 2015 RANK	FY 2014 RANK	FY 2015 SALES ($ MILLION)	FY 2015 R&D INVESTMENT ($ MILLION)	FY 2015 GROSS MARGIN
25	23	4,664.0	571.0	76.8%

Scope of R&D:
Activision develops and produces titles using a model in which a core group of creative, production and technical professionals, in coordination with our marketing, finance and other departments, has responsibility for the entire development and production process, including the supervision and coordination of internal and external resources. This team assembles the necessary creative elements to complete and release content through, where appropriate, outside programmers, artists, animators, scriptwriters, musicians and songwriters, sound effects and special effects experts, and sound and video studios.

Chief Executive Officer . . . . Bobby Kotick
Industry . . . . Software
Headquarters. . . . . Santa Monica, CA
Website . . . . www.activisionblizzard.com
Years in RQ50 Index 7

---

26 EXXON MOBIL CORP XOM

FY 2015 RANK	FY 2014 RANK	FY 2015 SALES ($ MILLION)	FY 2015 R&D INVESTMENT ($ MILLION)	FY 2015 GROSS MARGIN
26	31	236,810.0	971.0	18.6%

Scope of R&D:
To maintain our competitive position, especially in light of the technological nature of our businesses and the need for continuous efficiency improvement, ExxonMobil's research and development organizations must be successful and able to adapt to a changing market and policy environment, including developing technologies to help reduce greenhouse gas emissions.

Chief Executive Officer . . . . Rex W. Tillerson
President of ExxonMobil Upstream Research Company . . . . Sara N. Ortwein
Industry . . . . Petroleum
Headquarters. . . . . Irving, TX
Website . . . . www.corporate.exxonmobil.com
Years in RQ50 Index 18

27 XILINX INC XLNX

FY 2015 RANK	FY 2014 RANK	FY 2015 SALES ($ MILLION)	FY 2015 R&D INVESTMENT ($ MILLION)	FY 2015 GROSS MARGIN
27	30	2,213.9	525.7	71.9%

Scope of R&D:
Our research and development (R&D) activities are primarily directed towards the design of new ICs and the development of new software design automation tools for hardware and embedded software, the design of logic IP, the adoption of advanced semiconductor manufacturing processes for ongoing cost reductions, performance and signal integrity improvements and lowering PLD power consumption.

Chief Executive Officer . . . . .Moshe Gavrielov
Senior Vice President & Chief Technology Officer . . .Ivo Bolsens
Industry . . . . .Semiconductors
Headquarters. . . . . .San Jose, CA
Website . . . . .www.xilinx.com
Years in RQ50 Index 11

28 FIRST SOLAR INC FSLR

FY 2015 RANK	FY 2014 RANK	FY 2015 SALES ($ MILLION)	FY 2015 R&D INVESTMENT ($ MILLION)	FY 2015 GROSS MARGIN
28	20	3,579.0	144.0	32.9%

Scope of R&D:
We continue to devote substantial resources to research and development with the primary objective of lowering the lifecycle cost of electricity generated by our PV solar power systems. We conduct our research and development activities primarily in the United States. Within our components business, we focus our research and development activities on, among other areas, continuing to increase the conversion efficiency and energy yield of our solar modules and continuously improving durability and manufacturing efficiencies, including throughput improvement, volume ramp, and material cost reduction.

Chief Executive Officer . . . . Mark Widmar
Chief Technology Officer . . . Raffi Garabedian
Industry . . . . Semiconductors
Headquarters. . . . . Tempe, AZ
Website . . . . www.firstsolar.com
Years in RQ50 Index 4

29 NETFLIX INC NFLX

FY 2015 RANK	FY 2014 RANK	FY 2015 SALES ($ MILLION)	FY 2015 R&D INVESTMENT ($ MILLION)	FY 2015 GROSS MARGIN
29	39	6,779.5	472.3	84.6%

Scope of R&D:
Technology and development expenses consist of payroll and related costs incurred in making improvements to our service offerings, including testing, maintaining and modifying our user interface, our recommendation, merchandising and streaming delivery technology and infrastructure. Technology and development expenses also include costs associated with computer hardware and software.

Chief Executive Officer . . . . Reed Hastings
Industry . . . . Video Rental
Headquarters. . . . . Los Gatos, CA
Website . . . . www.netflix.com
Years in RQ50 Index 3

30 BIOGEN INC BIIB

FY 2015 RANK	FY 2014 RANK	FY 2015 SALES ($ MILLION)	FY 2015 R&D INVESTMENT ($ MILLION)	FY 2015 GROSS MARGIN
30	42	10,763.8	1,909.6	90.5%

Scope of R&D:
Our research efforts are focused on better understanding the underlying biology of diseases so we can discover and deliver treatments that have the potential to make a real difference in the lives of patients with high unmet medical needs. By applying our expertise in biologics and our growing capabilities in small molecule, antisense, gene therapy, gene editing and other technologies, we target specific medical needs where we believe new or better treatments are needed.

Chief Executive Officer . . . . George A. Scangos, PhD
Executive Vice President and Chief Scientific Officer. . . . Spyros Artavanis-Tsakonas, PhD
Industry . . . . Biological Products
Headquarters. . . . . Cambridge, MA
Website . . . . www.biogen.com
Years in RQ50 Index 9

31 QLOGIC CORP QLGC

FY 2015 RANK	FY 2014 RANK	FY 2015 SALES ($ MILLION)	FY 2015 R&D INVESTMENT ($ MILLION)	FY 2015 GROSS MARGIN
31		458.9	144.3	67.6%

Scope of R&D:
Our industry is subject to rapid, regular and sometimes unpredictable technological change. Our ability to compete depends upon our ability to continually design, develop and introduce new products that take advantage of market opportunities and address emerging standards. Our strategy is to leverage our substantial base of architectural, systems and engineering expertise to address a broad range of server and storage networking solutions. We are engaged in the design and development of ASICs and adapters that are primarily based on one or more of Fibre Channel, iSCSI, FCoE and Ethernet technologies.

Chief Executive Officer . . . . Prasad Rampalli
Chief Technology Officer . . . Bala Ganeshan, PhD
Industry . . . . Computer Equipment
Headquarters. . . . . Aliso Viejo, CA
Website . . . . www.qlogic.com
Years in RQ50 Index 10

32 ROVI CORP

ROVI

FY 2015 RANK	FY 2014 RANK	FY 2015 SALES ($ MILLION)	FY 2015 R&D INVESTMENT ($ MILLION)	FY 2015 GROSS MARGIN
32	18	526.3	108.7	80.4%

Scope of R&D:
Our internal research and development efforts are focused on developing enhancements to existing products and new applications for our current technologies.

Chief Executive Officer . . . . Thomas Carson
Industry . . . . Patent Owner
Headquarters. . . . . San Carlos, CA
Website . . . . www.rovicorp.com
Years in RQ50 Index 4

33 F5 NETWORKS INC

FFIV

FY 2015 RANK	FY 2014 RANK	FY 2015 SALES ($ MILLION)	FY 2015 R&D INVESTMENT ($ MILLION)	FY 2015 GROSS MARGIN
33	33	1,919.8	263.8	84.8%

Scope of R&D:
Our success depends on our timely development of new products and features, market acceptance of new product offerings and proper management of the timing of the life cycle of our products.

Chief Executive Officer . . . . John McAdam
Executive Vice President of Product Development and Chief Technical Officer . . . Karl Triebes
Industry . . . . System Design
Headquarters. . . . . Seattle, WA
Website www.f5.com
Years in RQ50 Index 7

34 CIENA CORP

CIEN

FY 2015 RANK	FY 2014 RANK	FY 2015 SALES ($ MILLION)	FY 2015 R&D INVESTMENT ($ MILLION)	FY 2015 GROSS MARGIN
34	44	2,445.7	401.2	46.8%

Scope of R&D:
Our research and development efforts seek to extend our existing technologies, including our WaveLogic coherent optical processor for 200G and 400G optical transport, and to introduce terabit per second and greater transmission speeds. We are also focused on expanding high-capacity service delivery capabilities in our Packet Networking and Converged Packet Optical products for metro networks, data center interconnectivity and WAN applications. Separately, we are increasing the scale, density and capability of our packet offerings, reducing power and space requirements, and enabling NFV capabilities for applications in metro networks, user aggregation and data center connectivity. We are also focused on increasing software programmability of networks and enabling network operators to automate and accelerate the creation and delivery of new, cloud-based services. These efforts include investments in our Blue Planet software platform—which is designed to automate, orchestrate, and manage physical network resources and virtualized services across data centers and the WAN—and its integration across our portfolio and with additional third party network resources.

Chief Executive Officer . . . . .Gary B. Smith
Senior Vice President and Chief Technology Officer. .Stephen B. Alexander
Industry . . . . .Communications Equipment
Headquarters. . . . . .Hanover, MD
Website . . . . .www.ciena.com
Years in RQ50 Index 5

35 IAC/INTERACTIVECORP

IAC

FY 2015 RANK	FY 2014 RANK	FY 2015 SALES ($ MILLION)	FY 2015 R&D INVESTMENT ($ MILLION)	FY 2015 GROSS MARGIN
35	37	3,230.9	160.5	75.9%

Scope of R&D:
Our success depends, in part, on our ability to continue to introduce new and enhanced content, products and services in response to evolving trends and technologies and that otherwise resonate with our users and customers.

Chief Executive Officer . . . . Joey Levin
Industry . . . . Computer Services
Headquarters. . . . . New York, NY
Website . . . . www.iac.com
Years in RQ50 Index 2

36 QUALCOMM INC

QCOM

FY 2015 RANK	FY 2014 RANK	FY 2015 SALES ($ MILLION)	FY 2015 R&D INVESTMENT ($ MILLION)	FY 2015 GROSS MARGIN
36	50	25,281.0	5,477.0	63.8%

Scope of R&D:
We expect to continue to invest in research and development in a variety of ways in an effort to extend the demand for our products and services, including continuing the development of CDMA, OFDMA and other technologies, developing alternative technologies for certain specialized applications, participating in the formulation of new voice and data communication standards and technologies and assisting in deploying digital voice and data communications networks around the world.

Chief Executive Officer . . . . Steve Mollenkopf
Executive Vice President, Qualcomm Technologies, Inc. and Chief Technology Officer. . . Matthew S. Grob
Industry . . . . Semiconductors
Headquarters. . . . . San Diego, CA
Website . . . . www.qualcomm.com
Years in RQ50 Index 11

37 ALPHABET INC

GOOGL

FY 2015 RANK	FY 2014 RANK	FY 2015 SALES ($ MILLION)	FY 2015 R&D INVESTMENT ($ MILLION)	FY 2015 GROSS MARGIN
37	35	74,989.0	9,832.0	69.1%

Scope of R&D:
We continue to invest in our existing products and services, including search and advertising, as well as developing new products and services through research and product development. We often release early-stage products. We then use data and user feedback to decide if and how to invest further in those products.

Chief Executive Officer . . . . Larry Page
Industry . . . . Computer Services
Headquarters. . . . . Mountain View, CA
Website . . . . www.google.com
Years in RQ50 Index 2

38 EBAY INC

EBAY

FY 2015 RANK	FY 2014 RANK	FY 2015 SALES ($ MILLION)	FY 2015 R&D INVESTMENT ($ MILLION)	FY 2015 GROSS MARGIN
38		8,592.0	2,000.0	86.8%

Scope of R&D:
eBay Inc.'s platforms utilize a combination of proprietary technologies and services as well as technologies and services provided by others. We have developed intuitive user interfaces, buyer and seller tools and transaction processing, database and network applications that help enable our users to reliably and securely complete transactions on our sites. Our technology infrastructure simplifies the storage and processing of large amounts of data, eases the deployment and operation of large-scale global products and services and automates much of the administration of large-scale clusters of computers. Our infrastructure has been designed around industry-standard architectures to reduce downtime in the event of outages or catastrophic occurrences. We strive to continually improve our technology to enhance the buyer and seller experience and to increase efficiency, scalability and security.

Chief Executive Officer . . . . Devin Wenig
SVp, Chief Technology Officer . . . Steve Fisher
Industry . . . . Computer Services
Headquarters. . . . . San Jose, CA
Website . . . . www.ebayinc.com
Years in RQ50 Index 4

---

39 AMGEN INC

AMGN

FY 2015 RANK	FY 2014 RANK	FY 2015 SALES ($ MILLION)	FY 2015 R&D INVESTMENT ($ MILLION)	FY 2015 GROSS MARGIN
39		21,662.0	4,297.0	90.5%

Scope of R&D:
We focus our R&D on novel human therapeutics for the treatment of serious illness in the areas of oncology/hematology, cardiovascular disease, inflammation, bone health, nephrology and neuroscience. We take a modality-independent approach to R&D with a focus on biologics. Our discovery research programs may therefore yield targets that lead to the development of human therapeutics delivered as large molecules, small molecules, or other combination or new modalities.

Chief Executive Officer . . . . Robert A. Bradway
Executive Vice President, Research and Development . . Sean E. Harper
Industry . . . . Biological Products
Headquarters. . . . . Thousand Oaks, CA
Website . . . . www.amgen.com
Years in RQ50 Index 13

---

40 POLARIS INDUSTRIES INC

PII

FY 2015 RANK	FY 2014 RANK	FY 2015 SALES ($ MILLION)	FY 2015 R&D INVESTMENT ($ MILLION)	FY 2015 GROSS MARGIN
40	40	4,719.3	148.5	30.9%

Scope of R&D:
We have approximately 750 employees who are engaged in the development and testing of existing products and research and development of new products and improved production techniques.

Chief Executive Officer . . . . Scott W. Wine
Vice President - Chief Technical Officer . . . Stephen J. Kemp
Industry . . . . Transportation Equipment
Headquarters. . . . . Medina, MN
Website . . . . www.polaris.com
Years in RQ50 Index 4

---

41 APPLE INC

AAPL

FY 2015 RANK	FY 2014 RANK	FY 2015 SALES ($ MILLION)	FY 2015 R&D INVESTMENT ($ MILLION)	FY 2015 GROSS MARGIN
41		233,715.0	6,041.0	44.6%

Scope of R&D:
Because the industries in which the Company competes are characterized by rapid technological advances, the Company's ability to compete successfully depends heavily upon its ability to ensure a continual and timely flow of competitive products, services and technologies to the marketplace. The Company continues to develop new technologies to enhance existing products and to expand the range of its product offerings through R&D, licensing of intellectual property and acquisition of third-party businesses and technology.

Chief Executive Officer . . . . Tim Cook
Senior Vice President, Hardware Technologies . . Johny Srouji
Industry . . . . Communications Equipment
Headquarters. . . . . Cupertino, CA
Website . . . . www.apple.com
Years in RQ50 Index 17

---

42 ZIMMER BIOMET HOLDINGS INC ZBH

FY 2015 RANK	FY 2014 RANK	FY 2015 SALES ($ MILLION)	FY 2015 R&D INVESTMENT ($ MILLION)	FY 2015 GROSS MARGIN
42		5,997.8	188.3	82.0%

Scope of R&D:
We have extensive research and development activities to develop new surgical techniques, materials, biologics and product designs. The research and development teams work closely with our strategic brand marketing function. The rapid commercialization of innovative new materials, biologics products, implant and instrument designs and surgical techniques remains one of our core strategies and continues to be an important driver of sales growth.

Chief Executive Officer . . . . David Dvorak
Industry . . . . Medical Equipment
Headquarters. . . . . Warsaw, IN
Website . . . . www.zimmerbiomet.com
Years in RQ50 Index 1

43 CONOCOPHILLIPS COP

FY 2015 RANK	FY 2014 RANK	FY 2015 SALES ($ MILLION)	FY 2015 R&D INVESTMENT ($ MILLION)	FY 2015 GROSS MARGIN
43	41	29,564.0	263.0	31.9%

Scope of R&D:
ConocoPhillips invests in R&D in order to reduce cost of supply from new developments, increase margins on production from existing fields and implement sustainability related technologies. Innovation focus areas include: developing unconventional reservoirs, improving seismic imaging, producing oil sands economically with fewer emissions, improving the economic efficiency of our company's proprietary LNG technology, reducing the time required to drill wells, and developing data analytics solutions.

Chief Executive Officer . . . . Ryan Lance
Executive Vice President, Technology and Projects . . . . Alan J. Hirshberg
Industry . . . . Petroleum
Headquarters. . . . . Houston, TX
Website . . . . www.conocophillips.com
Years in RQ50 Index 15

44 IONIS PHARMACEUTICALS INC IONS

FY 2015 RANK	FY 2014 RANK	FY 2015 SALES ($ MILLION)	FY 2015 R&D INVESTMENT ($ MILLION)	FY 2015 GROSS MARGIN
44		283.7	156.2	46.6%

Scope of R&D:
We are leaders in discovering and developing RNA-targeted therapeutics. We have created an efficient and broadly applicable drug discovery platform. Using this platform, we have developed a large, diverse and advanced pipeline of potentially first-in-class and/or best-in-class drugs that we believe can provide high value for patients with significant unmet medical needs.

Chief Executive Officer . . . . .Stanley T. Crooke, MD, PhD
Senior Vice President, Research . . .C. Frank Bennett, PhD
Industry . . . . .Pharmaceuticals
Headquarters. . . . . .Carlsbad, CA
Website . . . . .www.ionispharma.com
Years in RQ50 Index 1

45 APPLIED MATERIALS INC AMAT

FY 2015 RANK	FY 2014 RANK	FY 2015 SALES ($ MILLION)	FY 2015 R&D INVESTMENT ($ MILLION)	FY 2015 GROSS MARGIN
45		9,659.0	1,428.0	44.9%

Scope of R&D:
Applied's long-term growth strategy requires continued development of new products, including products that enable expansion into new markets. Applied's significant investments in research, development and engineering (RD&E) must generally enable it to deliver new products and technologies before the emergence of strong demand, thus allowing customers to incorporate these products into their manufacturing plans during early-stage technology selection. Applied works closely with its global customers to design systems and processes that meet their planned technical and production requirements.

Chief Executive Officer . . . . Gary E. Dickerson
Senior Vice President,
Chief Technology Officer . . Omkaram Nalamasu, PhD
Industry . . . . Machinery
Headquarters. . . . . Santa Clara, CA
Website . . . . www.appliedmaterials.com
Years in RQ50 Index 7

46 DOW CHEMICAL

DOW

FY 2015 RANK	FY 2014 RANK	FY 2015 SALES ($ MILLION)	FY 2015 R&D INVESTMENT ($ MILLION)	FY 2015 GROSS MARGIN
46	49	48,778.0	1,647.0	26.8%

Scope of R&D:
Dow research is metrics driven, customer focused and open to collaboration. It is through the combination of these principles that we earn the support of our shareholders for funding innovation efforts. Dow deploys metrics to analyze performance and forward-looking predictions to manage the R&D portfolio. Earnings from new products, patent advantaged sales and the NPV of the innovation portfolio are all used to manage the R&D investment. R&D productivity takes several forms. Our significant investment in high-throughput research enables us to run more experiments, optimizing products and processes faster and bringing products to market rapidly. We strive to "learn fast", when that is the correct decision, moving resources to projects that will deliver the most value. Dow collaborates both to extend our lab bench and to gain a seat at the design table with customers. We scout the globe for new innovations, collaborating to meet our customer needs.

Chief Executive Officer . . . . Andrew N. Liveris
Chief Technology Officer . . . A. N. Sreeram
Industry . . . . Plastics Materials
Headquarters. . . . . Midland, MI
Website . . . . www.dow.com
Years in RQ50 Index 15

47 HUNTSMAN CORP

HUN

FY 2015 RANK	FY 2014 RANK	FY 2015 SALES ($ MILLION)	FY 2015 R&D INVESTMENT ($ MILLION)	FY 2015 GROSS MARGIN
47		10,299.0	158.0	21.7%

Scope of R&D:
We support our business with a major commitment to research and development, technical services and process engineering improvement.

Chief Executive Officer . . . . Peter R. Huntsman
Chief Information Officer . . . Delaney Bellinger
Industry . . . . Industrial Chemicals
Headquarters. . . . . The Woodlands, TX
Website . . . . www.huntsman.com
Years in RQ50 Index 2

48 VERIFONE SYSTEMS INC

PAY

FY 2015 RANK	FY 2014 RANK	FY 2015 SALES ($ MILLION)	FY 2015 R&D INVESTMENT ($ MILLION)	FY 2015 GROSS MARGIN
48	46	2,000.5	203.7	45.2%

Scope of R&D:
Our R&D activities include design and development of our hardware products and unique operating systems, development of new solutions and applications, attaining applicable certifications and approvals required for our products and solutions, and ensuring compatibility and interoperability between our solutions and those of third parties. We work with our clients to develop system solutions that address existing and anticipated end-user needs.

Chief Executive Officer . . . . Paul S. Galant
Chief Technology Officer . . . Christophe Job
Industry . . . . Accounting Machines
Headquarters. . . . . San Jose, CA
Website . . . . www.verifone.com
Years in RQ50 Index 3

49 DOLBY LABORATORIES INC

DLB

FY 2015 RANK	FY 2014 RANK	FY 2015 SALES ($ MILLION)	FY 2015 R&D INVESTMENT ($ MILLION)	FY 2015 GROSS MARGIN
49	37	970.6	183.1	97.4%

Scope of R&D:
We have historically focused the majority of our R&D resources on audio technologies. In recent years, we have expanded our efforts to identify and develop new technologies. Beyond the strong audio platform we have created, we announced two new platforms during fiscal 2014—Voice and Imaging. Each of these platforms can support many offerings and we anticipate bringing new products to market using these platforms in the future.

Chief Executive Officer . . . . Kevin Yeaman
Senior Vice President, Advanced Technology Group . . Steven E. Forshay
Industry . . . . Patent Owner
Headquarters. . . . . San Francisco, CA
Website . . . . www.dolby.com
Years in RQ50 Index 5

50 MICROSEMI CORP MSCC

FY 2015 RANK	FY 2014 RANK	FY 2015 SALES ($ MILLION)	FY 2015 R&D INVESTMENT ($ MILLION)	FY 2015 GROSS MARGIN
50		1,245.6	192.0	57.8%

Scope of R&D:
The principal focus of our research and development activities has been to improve processes and to develop new products that support the growth of our businesses. The spending on research and development was principally to develop new higher-margin application-specific products, including, among others, our process and core architecture development for next generation programmable products, higher power PoE solutions, the continued roadmap development of our industry-leading timing and synchronization products, our silicon germanium (SiGe) RF power amplifier solutions for wireless LAN applications, and the ongoing development of gallium nitride (GaN) and silicon carbide (SiC) power management and RF solutions.

Chief Executive Officer . . . . Jim Peterson
Chief Technology Officer . . . Jim Aralis
Industry . . . . Semiconductors
Headquarters. . . . . Aliso Viejo, CA
Website . . . . www.microsemi.com
Years in RQ50 Index 2

NOTES

PREFACE

1. Hyland, L.A. "Pat" (1993), *Call Me Pat*, The Donning Company.

CHAPTER 1

1. Jaruzelski, B., K. Schwartz, and V. Staack (2015), Strategy& Global Innovation 1000 Study, http://www.strategyand.pwc.com/innovation1000.
2. Schwartz, L., R. Miller, D. Plummer, and A. Fusfeld (2011), "Measuring the Effectiveness of R&D," *Research-Technology Management* 54(5): 29–36.
3. Ibid.
4. Cohen, W., R. Nelson, and J. Walsh (2000). "Protecting Their Intellectual Assets: Appropriability Conditions and Why U.S. Manufacturing Firms Patent (or Not)," *NBER Working Paper No. 7552*.
5. http://www.ncbi.nlm.nih.gov/pmc/articles/PMC1523369/.
6. Scherer, F.M., and D. Harhoff (2000), Technology Policy For a World of Skew-Distributed Outcomes, *Research Policy* 29, 559–566.
7. An existing trademark prevented me from using IQ outside the academic domain.
8. Lewis, M. (2004), *Moneyball: The Art of Winning an Unfair Game*, W. W. Norton & Company.
9. http://www.nsf.gov/awardsearch/showAward?AWD_ID=0965147.

CHAPTER 2

1. Shapiro, G. (2011), *How Innovation Will Restore the American Dream*, Beaufort Books.
2. Schumpeter, J. (1942), *Capitalism, Socialism and Democracy*, Harper & Row.
3. Stevens and Burley (1997), "3,000 Raw Ideas = 1 Commercial Success!" *Research-Technology Management*, 40(3): 16–27.

4. Kengatharan, M., "Freakonomics of Factors That Determine Failure of a Biotech Company," http://qb3.org/sites/qb3.org/files/QB3Podcast20120521_6.pdf.
5. Nelson, R. (1959), "The Simple Economics of Basic Scientific Research," *The Journal of Political Economy*, 67(3): 297–306.
6. https://www.justice.gov/atr/case-document/complaint-189.
7. Zenger, T. (2016), *Beyond Competitive Advantage*, HBS Press.
8. Zenger, T. (1994), "Explaining Organizational Diseconomies of Scale in R&D: The Allocation of Engineering Talent, Ideas, and Effort by Company Size," *Management Science*, 40 (6): 708–729.
9. Allen, T. (1977), *Managing the Flow of Technology*, MIT Press.
10. Liu, C. (2016), "Brokerage by Design; Formal Structure, Geographic Locations, and Crosscutting Ties," working paper.
11. Nelson, R. (1959), "The Simple Economics of Basic Scientific Research," *The Journal of Political Economy*, 67(3): 297–306.
12. https://www.nsf.gov/statistics/srvyindustry/about/brdis/.
13. http://spectrum.mit.edu/spring-2014/the-brilliance-of-basic-research.
14. https://www.acs.org/content/acs/en/education/whatischemistry/landmarks/carotherspolymers.html.
15. Cohen, W. (2010), "Fifty Years of Empirical Studies of Innovative Activity and Performance," in B. H. Hall and N. Rosenberg, eds., *Handbook of Economics of Innovation*, North Holland Elsevier.
16. Posen, H. (2005), *Three Essays on Innovation and the Impact of Capital Markets*, http://repository.upenn.edu/dissertations/AAI3197730.
17. Inc. staff (2003), "Brief Profiles of 2003 Inc. 500 Companies," *Inc. Magazine*, Fall: 34.
18. Klepper, S. (2002), "The Capabilities of New Firms and the Evolution of the US Automobile Industry," *Industrial and Corporate Change*, 11 (4): 645–666.
19. Bhaskarabhatla, A., and S. Klepper (2014), "Latent Submarket Dynamics and Industry Evolution: Lessons from the US Laser Industry," *Industrial and Corporate Change*, http://icc.oxfordjournals.org/content/early/2014/01/12/icc.dtt060.
20. Smith, D., and R. Alexander (1999), *Fumbling the Future: How Xerox Invented, Then Ignored, the First Personal Computer*, iUniverse.
21. Lerner, J. (1995), *Xerox Technology Ventures: March 1995*, Harvard Business Review Case #: 295127-PDF-ENG.

22. Anand, B. (2007), *Schibsted*, Harvard Business Review Case #: 707474-PDF-ENG.
23. Gompers, P. (2002), "Corporations and the Financing of Innovation: The Corporate Venturing Experience," *Economic Review*, Q4:1–17.

CHAPTER 3

1. Porter, M.E. (1980), *Competitive Strategy*, Free Press, New York.
2. Kim, W.C., and R. Mauborgne (2005), *Blue Ocean Strategy: How to Create Uncontested Market Space and Make Competition Irrelevant*, Harvard Business Review Press.
3. Porter, M.E. (1990), *Competitive Advantage of Nations*, Free Press, New York.
4. Hou, K., and D.T. Robinson (2006), "Industry Concentration and Average Stock Returns," *Journal of Finance*, 61(4): 1927–1956.
5. Schumpeter, J. (1942), *Capitalism, Socialism, and Democracy*, New York: Harper and Row.
6. Adner, R., and D. Levinthal (2001), "Demand Heterogeneity and Technology Evolution: Implications for Product and Process Innovation," *Management Science*, 47(5): 611–628.
7. http://www.virgin.com/AboutVirgin/WhatWeAreAbout as of November 1, 2016. The page has since been updated and no longer contains this quote.
8. Bower, J. L., and C. M. Christensen (1995), "Disruptive Technologies: Catching the Wave," *Harvard Business Review*, 73(1): 43–53.
9. "The Columnists," *WSJ Magazine*, July/August 2016, 18.

CHAPTER 4

1. Smith, A. (1776), *Wealth of Nations*.
2. Romer, P.A. (1990), "Endogenous Technological Change," *Journal of Political Economy*, 98(5): S71–S102.
3. Solow, R. (1957), "Technical Change and the Aggregate Production Function," *Review of Economics and Statistics*, 39(3): 312–320.
4. Thompson, P. (1996), "Technological Opportunity and the Growth of Knowledge: A Schumpeterian Approach to Measurement," *Journal of Evolutionary Economics*, 6: 77; Lentz, R., and D. Mortensen (2008), "An Empirical Model of Growth Through Product Innovation," *Econometrica*, 76(6): 1317–1373.

5. Jones, C. (1995), "R&D-Based Models of Economic Growth", *Journal of Political Economy*, 103(4): 759–784.
6. Gordon, R. (2016), *The Rise and Fall of American Growth*, Princeton University Press.
7. Hartmann, Myers, and Rosenbloom (2006), "Planning your Company's R&D Investment," *Research-Technology Management*, March–April, 27.

CHAPTER 5

1. Lauren Coleman-Lochner and Carol Hymowitz, "Lafley's CEO Encore at P&G Puts Rock Star Legacy at Risk: Retail," May 28, 2013.
2. Mansfield, E. (1981), "Composition of R and D Expenditures: Relationship to Size of Firm, Concentration, and Innovative Output," *The Review of Economics and Statistics*, 63(4): 610615.
3. Nelson, R.R. (1959), "The Simple Economics of Basic Scientific Research," *Journal of Political Economy*, 67, 297306.
4. http://csdd.tufts.edu/files/uploads/Tufts_CSDD_briefing_on_RD_cost_study_-_Nov_18,_2014..pdf.
5. Ibid.
6. Stevens, G., and J. Burley (1997), "3,000 Raw Ideas = 1 Commercial Success!," *Research-Technology Management*, 3: 16–27.
7. Guler, I. (2003), "A Study of Decision Making, Capabilities and Performance in the Venture Capital Industry," Dissertations available from ProQuest. Paper AAI3095885.
8. Rosen, R. (1991), "Research and Development with Asymmetric Company Sizes," *The RAND Journal of Economics*, 22(3): 411–429.
9. http://web.stanford.edu/dept/SUL/sites/mac/parc.html.
10. Schawlow and Townes (1958), "Infrared and Optimal Masers," *Physical Review*.
11. https://www.aps.org/programs/outreach/history/historicsites/maiman.cfm.
12. Adner, R. (2012), *The Wide Lens*, portfolio.
13. Maimann, T. (2000), *The Laser Odyssey*, Laser Press.
14. http://moller.com.
15. Wu, B., and A.M. Knott (2006), "Entrepreneurial Risk and Market Entry," *Management Science*, 52(9): 1315–1330.
16. Elfenbein, D., A.M. Knott, and R. Croson, "Equity Stakes and Exit: An Experimental Approach to Decomposing Exit Delay," *Strategic Management Journal*, forthcoming.

17. Lerner, J. (1995), "Xerox Technology Ventures: March 1995," *Harvard Business Review*, Case Study 295127-PDF-ENG.
18. Bhaskarabhatla, A., and D. Hegde (2014), "An Organizational Perspective on Patenting and Open Innovation," *Organization Science*, 25(6): 1744–1763.

CHAPTER 6

1. Chesbrough, H. (2003), *Open Innovation: The New Imperative for Creating and Profiting from Technology*, Harvard Business Review Press.
2. Arora, A., W. Cohen, and J. Walsh (2014), "The Acquisition and Commercialization of Invention in American Manufacturing: Incidence and Impact," *NBER working paper* w20264.
3. Hippel, Eric von (1986), "Lead Users: A Source of Novel Product Concepts," *Management Science*: 791–805.
4. Kuan, J., D. Snow, and S. Helper (2014), "Supplier Innovation Strategy: Transactional Hazards and Innovation in the Automotive Supply Chain," working paper.
5. Boudreau, K., and K. Lakhani (2013), "Using the Crowd as an Innovation Partner," *Harvard Business Review*, 91(4): 61–69.
6. Levina, Fayard, and Gkeredakis (2014), "Organizational Impacts of Crowd Sourcing: What Happens with 'Not Invented Here' Ideas," working paper.
7. Oshri, I., and J. Kotlarsky (2011), "Innovation in Outsourcing: A Study of Client Expectations and Commitment," Warwick Business School working paper.
8. Morris, F., and B. Shackelford (2014), "Extramural R&D Funding by U.S.-Located Businesses Nears $30 Billion in 2011," *National Science Foundation, National Center for Science and Engineering Statistics*, NSF 14-314.
9. Weigelt, C. (2009), "The Impact of Outsourcing New Technologies on Integrative Capabilities and Performance," *Strategic Management Journal*, 30, 595–616.
10. Azoulay, P. (2004), "Capturing Knowledge Within and Across Company Boundaries: Evidence from Clinical Development," *American Economic Review*, 94(5): 1591–1612.
11. Bikhchandani, S., D. Hirshleifer, and I. Welch (1992), "A Theory of Fads, Fashion, Custom, and Cultural Change as Informational Cascades," *Journal of Political Economy*, 100(5): 992–1026.

12. Pisano, G., and W. Shih (2012), *Producing Prosperity: Why America Needs a Manufacturing Renaissance*, Harvard Business Review Press.

CHAPTER 7

1. Geisler, E. (2001), *Creating Value with Science and Technology*, Praeger.
2. Knott, A. (1994), *GM Hughes Electronics*, unpublished teaching case.
3. Ibid.
4. Coleman-Lochner, L., and C. Hymovitz (2012), "P&G's 1,000 PhDs Not Enough to Crank Up New Blockbusters," *Bloomberg*, September 12, 2012.
5. Ibid.
6. Brown, B., and S. Anthony (2011), "How P&G Tripled Its Innovation Success Rate," *Harvard Business Review*, 89(6): 64–72.
7. Nickerson, J., and T. Zenger (2002), "Being Efficiently Fickle: A Dynamic Theory of Organizational Choice," *Organization Science* 13(5), 547–566.
8. Argyres, N., and B. Silverman (2004), "R&D, Organization Structure, and the Development of Corporate Technological Knowledge," *Strategic Management Journal*, 25(8–9): 929–958.
9. Arora, A., S. Belenzon, and L. Rios (2014), "Make, Buy, Organize: The Interplay Between Research, External Knowledge, and Company Structure," *Strategic Management Journal*, 35(3): 317–337.
10. http://www.forbes.com/sites/chunkamui/2011/10/17/five-dangerous-lessons-to-learn-from-steve-jobs/#68556e3d60da.

CHAPTER 8

1. http://blog.trade-radar.com/2009/07/how-to-analyze-tech-stocks-7-factors.html.
2. http://www.forbes.com/sites/innovatorsdna/2016/05/18/how-we-rank-the-most-innovative-mid-cap-companies-2016/#365ddd49d7e3.
3. Note that some companies in the Forbes 100 aren't in the figure because they either don't conduct R&D or aren't traded on U.S. exchanges (the source of the RQ data).
4. Damodaran, A. (2009), "Valuing Companies with Intangible Assets," working paper.

5. Financial Accounting Standards Board (1974), *Statement of Financial Accounting Standards No. 2: Accounting for Research and Development Costs.*
6. As quoted in Hagerty, J. (2013), "50,000 Products but 3M Still Searching for Growth: CEO Thulin Sticks with Long-Term Research," *Wall Street Journal*, November 18, 2013.
7. It may seem paradoxical that tying CEO compensation to stock price would favor R&D given my argument that the market doesn't know how to value R&D. However, the market ultimately values the R&D once it shows up in profits. CEOs with a long-term perspective expect to stay long enough to capture that gain.
8. While these returns aren't risk adjusted, the beta of the RQ50 is comparable to that of the S&P 500.
9. Knott, A.M., and C. Vieregger (2016), "The Puzzle of Market Value from R&D," http://ssrn.com/abstract=2382885.
10. Cooper, M., A.M. Knott, and W. Yang (2016), "Measuring Innovation," http://ssrn.com/abstract=2631655.
11. Note this is a shorthand valuation that assumes costs are changing in proportion to revenues.

CHAPTER 9

1. Jones, C. (1995), "R&D-Based Models of Economic Growth," *Journal of Political Economy*, 103(4): 759–784.
2. Jones, B. (2009), "The Burden of Knowledge and the Death of the Renaissance Man," *The Review of Economic Studies*, 76 (1): 283–317.
3. http://www.mapsofworld.com/world-top-ten/world-top-ten-personal-computers-users-map.html.
4. http://www.nytimes.com/1984/11/23/business/typewriters-of-electronic-era.html.
5. Chandler, A. (1962), *Strategy and Structure: Chapters in the History of the American Industrial Enterprise*, MIT Press.
6. Jin, G., and P. Leslie (2005), "The Case in Support of Restaurant Hygiene Grade Cards," Choices, *Agricultural and Applied Economics Association*, 20(2).
7. Cutler, D., R. Huckman, and M. Landrum (2004), "The Role of Information in Medical Markets: An Analysis of Publicly Reported Outcomes in Cardiac Surgery," *American Economic Review Papers and Proceedings*, 94(2): 342–346.

8. Song, H., A. Tucker, K. Murrell, and D. Vinson (2015), "Public Relative Performance Feedback in Complex Service Systems: Improving Productivity Through the Adoption of Best Practices," working paper.
9. Taylor, F. (1911), *The Principles of Scientific Management*, Harper & Brothers.
10. http://www.trimble.com/Corporate/About_History.aspx.

CHAPTER 10

1. Hall, B., J. Mairesse, and P. Mohnen (2010), "Measuring the Returns to R&D," *Handbook of the Economics of Innovation*, B. H. Hall and N. Rosenberg (eds.), North-Holland.
2. Cohen, W.M., R.R. Nelson, and J.P. Walsh (2000), "Protecting Their Intellectual Assets: Appropriability Conditions and Why U.S. Manufacturing Companies Patent (Or Not)," NBER Working Paper 7552.
3. Abrams, D.S., U. Akcigit, and J. Popadak (2013), "Patent Value and Citations: Creative Destruction or Strategic Disruption?," NBER Working Paper 19647.
4. For a summary, see Redding, S. (2010), "Theories of Heterogeneous Firms and Trade," NBER Working Paper 16562.
5. If the R&D decision is made at the same time as those for other inputs, then all inputs should be jointly optimized. We make the simplifying assumption that because the benefits of R&D are lagged, it is chosen a year prior to the decisions for other inputs.
6. Thompson, P. (1996), "Technological Opportunity and the Growth of Knowledge: A Schumpeterian Approach to Measurement," *Journal of Evolutionary Economics*, 6(1), 77–97; Lentz, R., and D. Mortensen (2008), "An Empirical Model of Growth Through Product Innovation," *Econometrica*, 76(6), 1317–1373.
7. Knott, A.M., and C. Vieregger (2014), "An Empirical Test of Endogenous Company Growth," SSRN working paper, http://ssrn.com/abstract=2382885.

INDEX

ABOUT THE AUTHOR

Anne Marie Knott, PhD, is Professor of Strategy at the Olin Business School of Washington University, where she has been on faculty since 2005. Prior to her appointment at Olin, she was Assistant Professor of Management at the Wharton School of the University of Pennsylvania.

Prior to her career as an academic, Professor Knott worked at Hughes Aircraft Company developing missile guidance systems. This background both motivates and informs her research. Her primary research area is innovation—both large-scale R&D and entrepreneurship, which she views as complementary in contributing to economic growth. Her work on the organizational research quotient (RQ) was awarded two grants from the National Science Foundation. She has supervised over 100 consulting projects across a broad range of industries and strategic issues. For the past five years, however, her consulting work has focused on use of the RQ measure to help firms improve management of their R&D.

Knott's research has been covered by CNBC, *Forbes*, *The Atlantic*, *Business Week*, and *The Wall Street Journal*. She has published numerous articles in *Harvard Business Review*, *Management Science*, *Organization Science*, the *Strategic Management Journal*, *Small Business Economics*, the *Journal of Economic Behavior and Organizations*, and *Research and Technology Management*.

Professor Knott received her PhD in management as well as her MBA from the UCLA Anderson Graduate School of Management.